Communications in Computer and Information Science   **624**

Commenced Publication in 2007
Founding and Former Series Editors:
Alfredo Cuzzocrea, Dominik Ślęzak, and Xiaokang Yang

Wanxiang Che · Qilong Han
Hongzhi Wang · Weipeng Jing
Shaoliang Peng · Junyu Lin
Guanglu Sun · Xianhua Song
Hongtao Song · Zeguang Lu (Eds.)

Social Computing

Second International Conference
of Young Computer Scientists,
Engineers and Educators, ICYCSEE 2016
Harbin, China, August 20–22, 2016
Proceedings, Part II

 Springer

Editors

Wanxiang Che
Harbin Institute of Technology
Harbin
China

Qilong Han
Harbin Engineering University
Harbin
China

Hongzhi Wang
Harbin Institute of Technology
Harbin
China

Weipeng Jing
Northeast Forestry University
Harbin
China

Shaoliang Peng
National University of Defense Technology
Changsha
China

Junyu Lin
Harbin Engineering University
Harbin
China

Guanglu Sun
Harbin University of Science
  and Technology
Harbin
China

Xianhua Song
Harbin University of Science
  and Technology
Harbin
China

Hongtao Song
Harbin Engineering University
Harbin
China

Zeguang Lu
Harbin Sea of Clouds and Computer
  Technology
Harbin
China

ISSN 1865-0929   ISSN 1865-0937   (electronic)
Communications in Computer and Information Science
ISBN 978-981-10-2097-1   ISBN 978-981-10-2098-8   (eBook)
DOI 10.1007/978-981-10-2098-8

Library of Congress Control Number: 2016945792

Printed on acid-free paper

This Springer imprint is published by Springer Nature
The registered company is Springer Science+Business Media Singapore Pte Ltd.

Preface

As the general and program co-chairs of the Second International Conference of Young Computer Scientists, Engineers and Educators 2016 (ICYCSEE 2016), it is our great pleasure to welcome you to the proceedings of the conference, which was held in Harbin, China, during August 20–22, 2016, hosted by Harbin Engineering University. The goal of this conference is to provide a forum for young computer scientists, engineers, and educators.

The call for papers of this year's conference attracted 338 paper submissions. After the hard work of the Program Committee, 91 papers were accepted to appear in the conference proceedings, with an acceptance rate of 27 %. The main theme of this conference was "Social Computing." The accepted papers cover a wide range of areas related to social computing such as: science and foundations for social computing, computation infrastructure for social computing, big data management analysis for social computing, evaluation methodologies for social computing and social media, intelligent computation for social computing, natural language processing techniques and culture analysis in social computing and social media, mobile social computing and social media, privacy and security in social computing and social media, public opinion analysis for social media, social modeling, social network analysis, user-generated content (wikis, blogs), and visualizing social interaction.

We would like to thank all the Program Committee members – 178 members from 84 institutes – for their hard work in completing the review tasks. Their collective efforts made it possible to attain quality reviews for all the submissions within a few weeks. Their diverse expertise in each individual research area helped us to create an exciting program for the conference. Their comments and advice helped the authors to improve the quality of their papers and gain deeper insights.

Our thanks also go to the authors and participants for their tremendous support in making the conference a success. Moreover, we thank Dr. Lanlan Chang and Jian Li from Springer, whose professional assistance was invaluable in the production of the proceedings.

Besides the technical program, this year ICYCSEE offered different experiences to the participants. We hope you enjoy the conference proceedings.

June 2016

Qilong Han
Wanxiang Che
Hongzhi Wang
Shoaling Peng
Junyu Lin

Organization

The Second International Conference of Young Computer Scientists, Engineers and Educators (ICYCSEE) 2016 (http://2016.icycsee.org) took place in Harbin, China, during August 2016 20–22, hosted by Harbin Engineering University.

ICYCSEE 2016 Steering Committee

Jianzhong, Li	Harbin Institute of Technology, China
Ting, Liu	Harbin Institute of Technology, China
Zhongbin, Su	Northeast Agricultural University, China
Guisheng, Yin	Harbin Engineering University, China

General Chairs

Qilong, Han	Harbin Engineering University, China
Wanxiang, Che	Harbin Institute of Technology, China

Program Chairs

Hongzhi, Wang	Harbin Institute of Technology, China
Shaoliang, Peng	National University of Defense Technology, China
Junyu, Lin	Harbin Engineering University, China

Organization Chairs

Hongtao, Song	Harbin Engineering University, China
Zeguang, Lu	Sea of Clouds and Computer Technology Services Ltd., China

Publication Chairs

Guanglu, Sun	Harbin University of Science and Technology, China
Zhaowen, Qiu	Northeast Forestry University, China

Publication Co-chairs

Weipeng, Jing	Northeast Forestry University, China
Xianhua, Song	Harbin University of Science and Technology, China

Education Chairs

Yingtao, Zhang	Harbin Institute of Technology, China
Zhongyang, Han	Heilongjiang Institute of Technology, China

Industrial Chair

Jiquan, Ma   Heilongjiang University, China

Demo Chairs

Changjian, Zhou   Northeast Agricultural University, China
Qi, Han   Harbin Institute of Technology, China

Panel Chairs

Haiwei, Pan   Harbin Engineering University, China
Hui, Gao   Harbin Huade University, China

Registration/Financial Chairs

Yong, Wang   Harbin Engineering University, China
Fa, Yue   Sea of Clouds and Computer Technology Services Ltd.,
                              China

Post/Expo Chair

Tingting, Chen   SuperMap Software Co., Ltd

ICYCSEE Steering Committee

Guanglu, Sun   Harbin University of Science and Technology, China
Hai, Jin   Huazhong University of Science and Technology,
                              China
Haoliang, Qi   Heilongjiang Institute of Technology, China
Hongzhi, Wang   Harbin Institute of Technology, China
Jiajun, Bu   Zhejiang University, China
Jian, Chen   PARATERA
Junyu, Lin   Harbin Engineering University, China
Liehuang, Zhu   Beijing Institute of Technology, China
Min, Zhu   Sichuan University, China
Qilong, Han   Harbin Engineering University, China
Shaoliang, Peng   National University of Defense Technology, China
Tao, Wang   Peking University, China
Tian, Feng   Institute of Software Chinese Academy of Sciences,
                              China
Wanqing, He   Qihoo360 Cloud Company
Wanxiang, Che   Harbin Institute of Technology, China
Weipeng, Jing   Northeast Forestry University, China
Xiaohui, Wei   Jilin University, China
Xiaoru, Yuan   Peking University, China

Xuebin, Chen	North China University of Science and Technology, China
Yanjuan, Sang	Beijing Gooagoo Technology Service Co., Ltd., China
Yiliang, Han	Engineering University of CAPF
Yingao, Li	Neuedu
Yinhe, Han	Institute of Computing Technology, Chinese Academy of Sciences, China
Yu, Yao	Northeastern University, China
Yunquan, Zhang	Institute of Computing Technology, Chinese Academy of Sciences, China
Zeguang, Lu	Harbin Sea of Clouds and Computer Technology Services Ltd., China
Zhaowen, Qiu	Northeast Forestry University, China
Zheng, Shan	The PLA Information Engineering University

Program Committee

Tian, Bai	Jilin University, China
Zhifeng, Bao	University of Tasmania, Australia
Jiajun, Bu	Zhejiang University, China
Zhipeng, Cai	Georgia State University, USA
Wanxiang, Che	Harbin Institute of Technology, China
Xuebin, Chen	Hebei United University, China
Wenliang, Chen	Soochow University, China
Siyao, Cheng	Harbin Institute of Technology, China
Dansong, Cheng	Harbin Institute of Technology, China
Yuan, Cheng	Harbin University of Science and Technology, China
Yan, Chu	Harbin Engineering University, China
Lei, Cui	Microsoft Research
Beiliang, Cui	Nanjing Tech University, China
Bin, Cui	Peking University, China
Jianrui, Ding	Harbin Institute of Technology, China
Minghui, Dong	Institute for Infocomm Research, Singapore
Xunli, Fan	Northwest University, China
Chunxiang, Fan	University of Ulm, Germany
Guangsheng, Feng	Harbin Engineering University, China
Yansong, Feng	University of Edinburgh, UK
Guohong, Fu	Heilongjiang University, China
Hui, Gao	Harbin Huade University, China
Shang, Gao	Jilin University, China
Jing, Gao	University at Buffalo, USA
Dianxuan, Gong	North China University of Science and Technology, China
Yi, Guan	Harbin Institute of Technology, China
Quanlong, Guan	Jinan University, China
Yuhang, Guo	Beijing Institute of Technology, China

Ma, Han	Georgia State University, USA
Qilong, Han	Harbin Engineering University, China
Zhongyuan, Han	Harbin Institute of Technology, China
Qi, Han	Harbin Institute of Technology, China
Xianpei, Han	Institute of Software, Chinese Academy of Sciences, China
Zhongjun, He	Baidu Inc.
Zhenying, He	Fudan University, China
Yu, Hong	Soochow University, China
Zhengang, Jiang	Changchun University of Science and Technology, China
Cheqing, Jin	East China Normal University, China
Peng, Jin	Peking University, China
Weipeng, Jing	Northeast Forestry University, China
Leilei, Kong	Heilongjiang Institute of Technology, China
Dapeng, Lang	Harbin Engineering University, China
Dan, Le	Harbin Institute of Technology, China
Mei, Li	China University of Geosciences (Beijing), China
Shuaicheng, Li	City University of Hong Kong, SAR China
Jie, Li	Harbin Institute of Technology, China
Zhixun, Li	Harbin Institute of Technology, China
Junbao, Li	Harbin Institute of Technology, China
Ao, Li	Harbin University of Science and Technology, China
Peng, Li	Institute of Information Engineering, CAS, China
Maoxi, Li	Jiangxi Normal University, China
Chenliang, Li	Wuhan University, China
Junyu, Lin	Harbin Engineering University, China
Xianmin, Liu	Harbin Institute of Technology, China
Shaohui, Liu	Harbin Institute of Technology, China
Ming, Liu	Harbin Institute of Technology, China
Xiaoguang, Liu	Nankai University, China
Hailong, Liu	Northwestern Polytechnical University, China
Chenguang, Liu	Samsung
Zhiyuan, Liu	Tsinghua University, China
Zeguang, Lu	Harbin Sea of Clouds and Computer Technology Services Ltd., China
Nan, Lu	Shenzhen University, China
Jizhou, Luo	Harbin Institute of Technology, China
Zhiyong, Luo	Harbin University of Science and Technology, China
Chengguo, Lv	Heilongjiang University, China
Yanjun, Ma	Baidu Inc.
Shuai, Ma	Beihang University, China
Jiquan, Ma	Heilongjiang University, China
Dapeng, Man	Harbin Engineering University, China
Tiezheng, Nie	Northeastern University, China
Haiwei, Pan	Harbin Engineering University, China

Liqiang, Pan	Harbin Institute of Technology, China
Wei, Pan	Northwestern Polytechnical University, China
Zhijuan, Peng	Nantong University, China
Shaoliang, Peng	National University of Defense Technology, China
Jian, Peng	Sichuan University, China
Yuwei, Peng	Wuhan University, China
Shaojie, Qiao	Southwest Jiaotong University, China
Zhijing, Qin	University of California, Irvine, USA
Xipeng, Qiu	Fudan University, China
Zhaowen, Qiu	Northeast Forestry University, China
Ying, Shan	Harbin Guangsha University, China
Juan, Shan	Pace University, USA
Bin, Shao	Microsoft Research Asia
Shengfei, Shi	Harbin Institute of Technology, China
Xianhua, Song	Harbin University of Science and Technology, China
Yangqiu, Song	West Virginia University, USA
Jie, Su	Harbin University of Science and Technology, China
Jinsong, Su	Xiamen University, China
Hailong, Sun	Beihang University, China
Xiaoling, Sun	Dalian University of Technology, China
Weiwei, Sun	Fudan University, China
Jianguo, Sun	Harbin Engineering University, China
Chengjie, Sun	Harbin Institute of Technology, China
Guanglu, Sun	Harbin University of Science and Technology, China
Dongpu, Sun	Harbin University of Science and Technology, China
Xu, Sun	Peking University, China
Buzhou, Tang	Harbin Institute of Technology of Shenzhen Graduate School, China
Jintao, Tang	National University of Defense Technology, China
Zhanyong, Tang	Northwest University, China
Jianhua, Tao	Chinese Academy of Sciences, China
Yongxin, Tong	Beihang University, China
Xifeng, Tong	Northeast Petroleum University
Zhiying, Tu	Harbin Institute of Technology, China
Zumin, Wang	Dalian University, China
Hongya, Wang	Donghua University, China
Haofen, Wang	East China University of Science and Technology, China
Xingmei, Wang	Harbin Engineering University, China
Hongzhi, Wang	Harbin Institute of Technology, China
Jinbao, Wang	Harbin Institute of Technology, China
Tiantian, Wang	Harbin Institute of Technology, China
Kechao, Wang	Harbin Institute of Technology, China
Wei, Wang	Institute of Software, Chinese Academy of Sciences, China
Botao, Wang	Northeastern University, China

Chaokun, Wang	Tsinghua University, China
Yanjie, Wei	Shenzhen Institutes of Advanced Technology, Chinese Academy of Sciences, China
Wei, Wei	Xi'an University of Technology, China
Xiuxiu, Wen	Harbin Engineering University, China
Xiaojun, Wen	Shenzhen Polytechnic, China
Xiangqian, Wu	Harbin Institute of Technology, China
Sai, Wu	Zhejiang University, China
Rui, Xia	Nanjing University of Science and Technology, China
Min, Xian	Utah State University, USA
Tong, Xiao	Northeastern University, China
Hui, Xie	Harbin Institute of Technology, China
Yu, Xin	Harbin Engineering University, China
Junchang, Xin	Northeastern University, China
Zheng, Xu	Harbin Institute of Technology, China
Jianliang, Xu	Hong Kong Baptist University, SAR China
Ying, Xu	Hunan University, China
Yongzeng, Xue	Harbin Institute of Technology, China
Ziye, Yan	Beijing Institute of Technology, China
Shaohong, Yan	North China University of Science and Technology, China
Hailu, Yang	Harbin University of Science and Technology, China
Xiaochun, Yang	Northeastern University, China
Yajun, Yang	Tianjin University, China
Xiaoyan, Yin	Northwest University, China
Shouyi, Yin	Tsinghua University, China
Xz, Yu	Harbin Institute of Technology, China
Haining, Yu	Harbin Institute of Technology, China
Zhengtao, Yu	Kunming University of Science and Technology, China
Xiaohui, Yu	Shandong University, China
Ye, Yuan	Northeastern Universtiy, China
Yingjun, Zhang	Beijing Jiaotong University, China
Rong, Zhang	East China Normal University, China
Liguo, Zhang	Harbin Engineering University, China
Zhiqiang, Zhang	Harbin Engineering University, China
Yingtao, Zhang	Harbin Institute of Technology, China
Delong, Zhang	Harbin Institute of Technology, China
Yu, Zhang	Harbin Institute of Technology, China
Ru, Zhang	Harbin University of Commerce, China
Meishan, Zhang	Heilongjiang University, China
Jiajun, Zhang	Institute of Automation Chinese Academy of Sciences, China
Rui, Zhang	Jilin University, China
Jian, Zhang	Northeast Forestry University
Xiao, Zhang	Renmin University of China, China
Hu, Zhang	Shanxi University, China

Contents – Part II

Industry Track

Demo Track

Contents – Part I

Education Track

Computer English Acquisition Environment Construction Based on Question Answering Technology

Fei Lang[1(✉)], Peipei Li[1], Jian Kang[2], and Guanglu Sun[3]

[1] School of Foreign Languages, Harbin University of Science and Technology,
Harbin, Heilongjiang, China
langfeihust@163.com
[2] Beijing Institute of Astronautical Systems Engineering, Beijing, China
[3] School of Computer Science and Technology,
Harbin University of Science and Technology, Harbin, Heilongjiang, China

Abstract. In-depth vocabulary knowledge plays a key role in L2 (second language) vocabulary acquisition, and computer English vocabulary is vital to academic performances of students majored in Computer Science. While many Chinese students cannot realize computer English vocabulary acquisition due to lacking input environment of computer English. To solve this problem, with techniques of keyword search, information retrieval, and text classification, to meet the diverse learning needs of students, it built the in-depth computer English vocabulary knowledge acquisition environment for the high frequency of computer English words, providing the depth of knowledge such as polysemy, synonyms, syntax, and pragmatics. The construction of in-depth computer English vocabulary acquisition environment helps future studies on computer English teaching applications.

Keywords: Computer English · In-depth vocabulary knowledge · Information retrieval · Text classification

1 Introduction

Since the 1970s of the last century, with the development of research on language acquisition, researchers gradually recognize the important role of vocabulary knowledge in language comprehension and output. From the microscopic analysis of the different dimensions of the single word [1–3] to the macroscopic description of the whole

This research is partly supported by the Ministry of Education's Humanities and Social Science Project No. 11YJC740048, Scientific planning issues of education in Heilongjiang Province No. GBC1211062, research fund for the program of new century excellent talents in Heilongjiang provincial university No. 1155-ncet-008 and the National Natural Science Foundation of China under grant Nos. 60903083, 61502123.

W. Che et al. (Eds.): ICYCSEE 2016, Part II, CCIS 624, pp. 3–9, 2016.
DOI: 10.1007/978-981-10-2098-8_1

vocabulary system [4–6]; from the emphasis on vocabulary accumulation to the exploration of characteristics of the vocabulary network construction, research on vocabulary acquisition develops fast.

As EFL (English as a foreign language) learners, Chinese students have little productive knowledge of English vocabulary; consequently, it is difficult for them to meet the needs of English communication, writing, and translation for the research and engineer in computer fields. The researchers found that there is still a big gap between Chinese students and native English speakers on productive vocabulary size, STTR(standard type-token ratio), the word length and other indicators, through the study on Chinese Learner English Corpus [7]. In fact, the most Chinese college students are able to identify and understand a large size of vocabulary, but can only employ a small part of it. Because their understanding of the majority of vocabulary still stays on the spelling, pronunciation, the core semantics, the basic syntax, and there is a lack of understanding of the depth of vocabulary knowledge such as polysemy, synonyms, syntax, pragmatics and so on.

The prevalence of the "plateau phenomenon" in Chinese learners' acquisition process of the depth of vocabulary knowledge is due to the lack of effective vocabulary teaching resources and methods [8]. To master the basic semantics of a word is not difficult, as students can reach the goal with words list. But to realize SLVA (special language vocabulary acquisition), specially for the students in computer science, relying solely on traditional teaching methods and textbooks in class is far from enough. Consequently, subjected to the traditional materials (target vocabulary scattered presentation, low repetition rate and a very small amount of contextual discourse), limited class time, individual differences and other factors, it is hard to accomplish SLVA in the traditional college English class.

2 Analysis

2.1 Research Questions

To tackle the problems of low English productive competence of EFL learners in China, and the inefficiency of computer English vocabulary teaching in traditional English class of Chinese colleges, it is to settle the questions that how to build depth of Computer English knowledge input environment based on question-answering technology.

2.2 Research Analysis

An L2 vocabulary test with high reliability and validity is the cornerstone to explore and build the L2 cognitive model. In the area of word associates test, Read's Word Associates Format (WAF) [9, 10] and Qian's Depth of Vocabulary Knowledge (DVK) [11] have been recognized by many test researchers [12].

Through building an acquisition system with functions of keyword-based search, retrieval and question-and-answer, it is to provide learners with depth knowledge information of a target English word such as polysemy, synonyms, collocation, syntax, and pragmatics etc., and to construct an open input environment for high-frequency computer English acquisition.

Moreover, although the existing corpus resources can provide English vocabulary teaching with the most commonly used part of speech, collocation and sentences of the target vocabulary, but for the purposes of the acquisition of the depth of vocabulary knowledge, language input with only basic lexical information is not enough.

So a new acquisition system is built based on question answering technology, such as information retrieval and text classification. Firstly, L2 database system will support a large amount of comprehensive L2 input (including: the target word polysemy, synonyms, syntactic knowledge, contextual discourse, pragmatic knowledge, etc.), which may create the necessary highly contextualized L2 input environment for L2 learners. Secondly, text classification will be utilized to classify different levels of computer reading materials, which makes learner read the suited technical materials in computer science. Finally, through the target word retrieval function and automatic difficulty rating function, it may extract related language input of the target word to meet the learners' actual English standard, which will not only achieve the goal of teaching students in accordance with their aptitude, but also enable students to freely choose topics or discourse types of the target word, which may stimulate learners' motivation.

3 Methods

3.1 Construction of L2 Database System

The effective acquisition of second language requires essential prerequisite for language input. Krashen puts forward the input hypothesis, holding that language input requires three prerequisites: comprehensible input; containing known language components; containing components prerequisite a little higher than known language. Ellis came up with a more specific principle, holding that language input is closely related with SLA in perspective of quantity, type, authenticity, optionality, extensibility, output ability, functionality, etc. Besides, the acquisition of profound meaning of words in the second language requires a proper learning order. The knowledge in different dimensions relate with each other.

The environment designed for SLA hence takes the following modes: First, pick a high-frequency word as a target word (we can add other words as relative words at the same time). And then do the searching job. Acquisition system searches the target word in the data bank of high-frequency words and renders the searching result in three ways. a. Focus on the synonym of the target language and render the information of the synonym; b. Focus on the lexis or syntax of the target language and render several phrases or sentences containing the target word or its relative word; c. Focus on the pragmatic level and render several contexts containing the target word or its relative word.

According to the multilevel classification function of discourse resource, acquisition system not only provides various sentences and discourses for learners automatically based on the learners' learning degree toward the target language, but also helps classify these discourses with different topics, to provide topic-related discourse setting.

Therefore, in regard to the acquisition need of high frequency words of profound knowledge in each dimension, the language input network is constructed in the following ways:

(1) The input of collocation knowledge of target word. (Common collocation and idioms; fixed and semi-fixed phrases and its variant; frequency and using degree of irregular form).

(2) The input of semantic items of target word (different meanings of target word in collocation, sentence, and discourse).

(3) The input of syntactic knowledge of target word (master the usage of the word in its syntactic level).

(4) The synonyms input of target word (synonym information; regarding synonyms as target word to get the further knowledge of synonyms).

(5) Using input of specific context and discourse to create acquisition environment of target word in pragmatic function (emotion; polysemy).

(6) General context and discourse input of target word helps create acquisition environment of connotation pragmatic function (culture, history, social meaning).

As we can see in Fig. 1, with key words searching and information retrieve, the profound knowledge of the target word "available" is fully performed. Therefore, learners have access to discourse that suitable for their English level with the help of multilevel classification function of discourse.

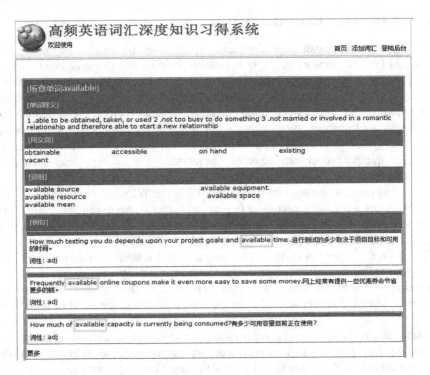

Fig. 1. In-depth knowledge acquisition system of high-frequency English vocabulary

3.2 Text Classification

The method that classifies the computer text is proposed here [13]. Based on the vocabulary in the last Section, the level of difficulty is defined as two levels, easy or difficult. SVM is utilized in classifier design.

SVM is a discriminative model which has strong theoretical basis and many successful application, which is utilized in a binary classification [14]. Assume that the training data and their labels are given as follows: $\{(x_1, y_1), (x_2, y_2), \ldots, (x_n, y_n)\}$, $x_i \in \mathbb{R}^d$, $y_i \in \{+1, -1\}$.

SVM gives a hyperplane that separates the training data by a maximal margin. The hyperplane is defined by the equation: $w \cdot x + b = 0$, where w is a coefficient vector, b is a scalar offset:

$$f(x) = w \cdot x = \sum_{i=1}^{n} w_i x_i \tag{1}$$

English texts are respectively labeled as -1 or 1, representing easy or difficult. Through Mercer kernel function $K(x_j, x_k) = \Phi(x_j) \cdot \Phi(x_k)$, e.g. linear, polynomial and RBF kernel, SVM allows mapping the original training data to a higher dimensional space in order to classify the data that is hard to separate in a low dimension space. Using Lagrange interpolation coefficients α_i, Formula (1) is transformed to solve a quadratic programming problem with linear constraints and its dual form with respect to vector α_i, $i = 1 \ldots n$. Therefore, the final discriminative function is:

$$f(x) = sign(w \cdot \Phi(x) + b) \tag{2}$$

SVM model optimizes the discriminative function with coefficients using all the training data based on sequential minimal optimization techniques.

The vector space in the SVMs model is abstracted as follows. The tagged texts are manually defined with two levels of difficulties. Words in the two levels of lexicons are extracted in the texts with corresponding level of difficult. The tagged texts are described as a vector of the occurrence frequency of words. These types of vectors are input of SVMs model.

3.3 Information Retrieval

Lucene is an open source high-performance, scalable, full-text indexing library. After ten years of development, Lucene has a large number of users and active development team, it was initially formed by the JAVA development, now has a C# and C++ and other transplant version.

The information retrieval function is realized by Lucene technology to build a local multi-form at search engine. Features include:

(1) On the txt, pdf, doc, xls, ppt and other popular text formats fast current index, the user can be simple and easy to operate front-end interface to enter the desired keyword query, the system can be fast in the file name and file content users need

to search for files and file names, file content and file size of the summary to the user, if users need to see the full content of the file search, you need to click on the file name can be, and in the search results to highlight the keywords, document summaries limited to words.

(2) It can be easily added to the lexicon of a new phrase in the Background maintenance system. And background maintenance personnel to establish and manage the index is very convenient, fast, just need to specify the source file directory can be specific, the system automatically searches for the source file and indexed.

(3) This system supports the fuzzy search feature, on the user's keyword for uncertain fuzzy matching; the results show a high similarity. So the search engine has a certain application.

4 Conclusion

The construction of in-depth vocabulary acquisition environment helps future studies on L2 vocabulary acquisition, autonomous learning mode of in-depth vocabulary knowledge, lexical testing, and other relevant teaching applications. Besides, it suggest new research method for exploring computer English vocabulary patterns by applying techniques of information retrieval and text classification to construct large scale in-depth vocabulary acquisition environment, which provides good practical conditions to future studies of computer English vocabulary acquisition, autonomous learning mode of in-depth vocabulary knowledge, computer English vocabulary testing, and other relevant teaching application.

References

1. Richards, J.: The Context of Language Teaching. CUP, Cambridge (1985)
2. Wesche, M., Paribkht, T.: Assessing vocabulary knowledge: depth vs. breadth. Can. Mod. Lang. Rev. **53**(1), 13–40 (1996)
3. Nation, P.: Learning Vocabulary in Another Language. CUP, Cambridge (2001)
4. Aitchison, J.: Words in the Mind: An Introduction to the Mental Lexicon. Blackwell, Oxford (2002)
5. Vanniarajan, S.: An interactive model of vocabulary acquisition. Appl. Linguist. **8**(2), 183–216 (1997)
6. Henriksen, B.: Three dimensions of vocabulary development. Stud. Second Lang. Acquisition **21**(2), 303–317 (1999)
7. Gui, S.C., Yang, H.Z.: Chinese Learner English Corpus. Shanghai Foreign Language Education Press, Shanghai (2005)
8. Dai, W.D., Ren, Q.M.: To Design Internet-based cognitive-psychological environment for vocabulary acquisition. Media Foreign Lang. Instr. **2**, 1–61 (2005)
9. Read, J.: Validating a test to measure depth of vocabulary knowledge. In: Kunnan, A. (ed.) Validation in Language Assessment, pp. 41–60. Lawrence Erlbaum, Mahwah (1998)
10. Read, J.: The development of a new measure of L2 vocabulary knowledge. Lang. Test. **10**(3), 355–371 (1993)
11. Qian, D., Schedl, M.: Evaluation of an in-depth vocabulary knowledge measure for assessing reading performance. Lang. Test. **21**(1), 28–52 (2004)

12. Schmitt, N., Ng, J.W.C., Garras, J.: The word associates format: validation evidence. Lang. Test. **28**(1), 105–126 (2011)
13. Lang, F., Sun, G.L., Shen, Y.W.: Text categorization in selecting authentic materials on tertiary level. In: The 6th International Forum on Strategic Technology, pp. 769–772 (2011)
14. Joachims, T.: Text categorization with support vector machines: learning with many relevant features. In: Proceedings of ECML-98, 10th European Conference on Machine Learning, Chemnitz, DE, 1998, pp. 137–142 (1998)

Computer English Teaching Based on WeChat

Fei Lang[1(✉)], Kexin Zhang[1], Peipei Li[1], and Guanglu Sun[2]

[1] School of Foreign Languages, Harbin University of Science and Technology,
Harbin, Heilongjiang, China
langfei@gmail.com
[2] School of Computer Science and Technology,
Harbin University of Science and Technology, Harbin, Heilongjiang, China

Abstract. Computer English plays a vital role in academic study for students majored in Computer Science. As the fast development of information technology, traditional English teaching methods cannot meet students' learning needs. WeChat is the most popular mobile application of social networks in China. Its features of real-time communication, interactivity and utilization of fragmentary time attract educators' attention to its application in teaching. This paper carried out an empirical study to explore applying WeChat in Computer English teaching. With WeChat, an English interaction environment was built, and interactive learning activities were implemented. The experimental results show teaching based on WeChat is effective to improve students' Computer English skills.

Keywords: Computer English · WeChat · Social networks · English teaching

1 Introduction

As the industry language in the field of computer and information technology, English has its own features which cannot be replaced by other languages. Whenever we learn the latest computer technology, or use the latest computer software and hardware products, computer professional English is inseparable. In order to meet this requirement, major colleges and universities opened computer professional English Course [1].

To help students to master professional vocabulary, read scientific articles, English literature, to improve the ability of listening, speaking and translating. For the major of computer students, when they surf the professional web site, access to academic papers, reference manuals, to participate in academic exchanges and international conferences, the use of professional computer English cannot be missed. Computer professional

This research is partly supported by the Ministry of Education's Humanities and Social Science Project No. 11YJC740048, Scientific planning issues of education in Heilongjiang Province No. GBC1211062, research fund for the program of new century excellent talents in Heilongjiang provincial university No. 1155-ncet-008 and the Natural Science Foundation of Heilongjiang Province under grant Nos. QC2015084, F201132.

W. Che et al. (Eds.): ICYCSEE 2016, Part II, CCIS 624, pp. 10–20, 2016.
DOI: 10.1007/978-981-10-2098-8_2

English has certain requests in listening and speaking, so the emphasis is to cultivate students' ability in using English in the professional field [2].

Currently, computer professional students' listening, speaking and translating training is not the teaching focus, classroom teaching only pay attention to training the students' vocabulary, sentence translation of materials. Most of the students say that the professional computer course is lack of real context, has few opportunities for performing the specialty English, and some students say even after a serious learning in computer professional English class the whole semester, their English actual ability is not improved in essence. They still cannot get rid of the embarrassing of "dumb English" and "deaf English".

WeChat is a more rapid instant messaging tool, has zero tariff, cross platform communication, display real-time input status and other functions. At present, WeChat is widely used in information communication, content publishing, interpersonal communication, media platforms, etc. WeChat as the currently most popular and the highest utilization rate social software with college students, can make full use of spare time to help the college students practice English listening and speaking ability anytime, anywhere. It can be said that it is a supplement to the classroom teaching. Students are the center of the WeChat learning platform. WeChat can effectively exert the subjective initiative of students. More importantly, WeChat learning platform provide a real communication environment for College English speaking and listening learning. Even if the introverted students is willing to participate in the learning activities, effectively exert their enthusiasm, initiative and creative [3].

Retrieval "WeChat, English" in CNKI Chinese Academic Journal, we obtain fewer than 30 related documents, the "WeChat and Computer professional English" is minority. This article attempts to combine computer English classes and WeChat, try to utilize WeChat to improve computers students' ability and skills in English listening, speaking and translation. In this paper, make the use of WeChat instant messaging, across space and other functions, for the aspect of the computer professional English listening, speaking and translating. Empirical research try to improve computer professional English practical application ability. Combined with professional computer knowledge, employment opportunities can be increased and improve employment rate can be improved.

2   Characteristics of Computer English Course

2.1   Characteristics of Computer English

2.1.1   Professional Term

The emergence of new terminology is growing. The emergence of some terms is associated with high-speed development of computer technology. For example: Internet, Intranet, Extranet, are all appeared with the development of Internet technology. With the development of network language, homophonic word, derivative and acronyms are constantly emerging. For instance IC means I see, ICQ means I see you, 3W means World Wide Web. As a result, we teachers must focus on the media, to understand dynamic information, timely grasp the new terms and new trade terms.

2.1.2 Thumbnail Words

As a result of the commands in the computer, advanced language statements in a computer to a certain space, starting from the principle of saving and succinct, to give full play to its effectiveness, computer is stored and displayed in the information, as far as possible in the form of a thumbnail. For instance, Rom—Read Only Memory, RAM—Random Access Memory, RISC—Reduced Instruction Set Computer, CISC—Complex Instruction Set Computer [4].

2.1.3 Compound New Words

In computer vocabulary, derivative (by prefix or suffix as word-building structure to form new words), Compound (usually in small rung and connection "a" word of words) in computer English accounts for a large proportion. Especially the article will often use some frequently used technical terms that has been given in the form of acronyms, which greatly increased the difficulty understanding of computer English. For example: file + based → file-based, front + user → front-user, paper + free → paper-free, User + centric → user-centric [5].

2.1.4 Derivative Words

The professional English vocabulary is formed with derived word-formation, mostly on the basis of the existing words. It is through the root with all kinds of prefixes and suffixes to form neologisms. In the affix, there is noun affix. Such as inter-, sub-, in-, tele-, micro-; adjective affix: im-, un-, -able, -al, -ing, -ed; verbal affixation: re-, under-, de-, -en, con-. Among them, the prefix adopted to form the words on the computer professional English accounts for a large proportion. There is typical derivatives such as multimedia, interface, microprocessor, subsystem.

2.1.5 Loan Words

The loan word refers to the use of public English and vocabulary in everyday life language to express the meaning of the professional. The loan word is from the manufacturer name, brand name, product code name, inventor of the name, place names, etc. It can also be turned into a professional meaning from general public English vocabulary. It is also given new meaning to the original words. For example:cache, semaphore, firewall, mailbomb, fitfall.

2.2 Features of Computer English Courses

2.2.1 Timeliness

Due to the rapid development of computer technology and new technologies emerged endlessly, New technical terms and jargon are also emerged consecutively, All of this makes computer English has a feature of timeliness, so the computer English teaching materials need to be updated and adjusted timely. In the process of practical teaching, In order to raise the timeliness level, Teachers should pay attention to the timeliness feature of the course and make a knowledge of new technical terms and jargon.

2.2.2 Utility

Computer and English are both belong to instrumental course, so computer English has a double instrumental features, we can understand the strength of its Utility from that Computer English can be regarded as a translation of Computer courses, The ultimate goal of offering a Computer course is make student to apply the English knowledge to Study and work.

2.2.3 Professional

We need to use and familiar with a lot evolved technical terms. There are many common English words has special meanings in professional field. For example, the word "Bus" means a large motor vehicle that carries passengers from one place to another, but its means "ROM, RAM" in professional field. Like "mouse" should be translated to "a device that is connected to computer",driver should be translated to "a computer program that controls a device such as a printer", memory should be translated to "the part of the computer where information is stored".We need to be familiar with the professional meanings of these words, for people who are unfamiliar with computer English, Computer technical terms will be a big problem. To understand every professional terms is the basic requirement for the teachers of computer English course.

3 Problems in English Teaching of Students Majored in Computer Science

3.1 Teacher Team Needs to Be Improved

Normally, the teacher who teaches computer professional course can't teach English well. This is also the common characteristics of other colleges and universities [4], A part of the computer professional English teaching course is English teachers, Another part is computer science graduates, If teachers of English majors have the professional course, his computer professional knowledge is insufficient, Most of them teach according to the teaching mode in public English teaching, In the course of teaching cannot complete the teaching content, Although computer professional teachers can explain the professional knowledge, they will be hampered by English grammar. Therefore, the key to the reform of computer professional English teaching is to have a high level of teachers, to have excellent academic leaders. They should not only have excellent knowledge of computer English, but also should have comprehensive application ability, strong organizational skills.

3.2 The Contents of the Textbooks Cannot Be Updated Timely

At present, many kinds of English courses of computer major in Colleges and universities of our country are quite different. These materials mainly include domestic and foreign related teaching materials and some original materials imported from abroad. In the choice of college computer professional English teaching materials should take into account the basic system and common operation of the computer, At the same time should keep up with the change of computer technology, the English in Teaching materials should be fluent, have Authentic language, The difficulty should be suitable for students to learn the actual situation, Should have a wide range of knowledge, Not only have already mastered professional knowledge, but also have not learned professional content, both have Professional basic knowledge and a higher level of professional content.

3.3 Lagging Teaching Methods

Compared with other courses, computer professional English is not professional required courses, coupled with teaching material content is usually reading the text, therefore teaching methods are relatively backward, and there is no innovation and improvement [6]. In traditional classroom teaching, teachers often use a blackboard and chalk, even with the use of multimedia course ware, network curriculum and advanced teaching means, they just simply list textbook content, did not get rid of "blackboard + chalk" cramming teaching methods. Teachers in teaching are generally in the interpretation based on new vocabulary, translate the textbook sentence, teaching methods are really single, lack of effective interaction with the students, so the students cannot fully exert their enthusiasm, students even feel boring, unattractive, they lose the interest in this course gradually, the students' lower learning enthusiasm has a serious impact on the students' learning effect.

3.4 Curriculum Setting Time is Short, the Students Do Not Pay Much Attention to the Course

Major universities basically make the computer English course the examine course, only 2 h a week to learn. Teachers and students do not have enough time to carry out bilateral teaching activities. Students understand the course is just to learn some professional vocabulary and technical English grammar. There are many versions of the current translation of computer books, so they think computer professional English does not have much relationship to their current study. Also, most students lack of research spirit, difficulties on the back, so the computer professional English learning is stagnant.

4 Feasibility Analysis of the WeChat Applied to Improve the Ability of English Listening and Speaking

4.1 Input Hypothesis

"Comprehensible language input" is a necessary condition for language acquisition. Comprehensible language means the language of the hearing or reading material can be understood by learners, the difficulty of these materials should be slightly higher than the knowledge of the language the learner has mastered [7].

4.2 Characteristics of the Comprehensible Input

Language input must be both interesting and associated, It not only refers to arouse the interest in learning but also relate to the real lives. Only a large amount of language input can arouse o language acquisition, which means a lot of related listening or reading.

4.3 Feasibility Analysis

Computer majors, English proficiency varies, but the actual course is "one size fits all" teaching mode, the difficulty of teaching materials is not suitable for all students, The distance between actual teaching content and practical application is far, which makes the students language input is not comprehensible, it cannot reach the effect of language acquisition, WeChat can realize group teaching, different levels of English students are grouped according to the actual level of listening, speaking.

Currently, the computer professional course in major colleges and universities is short, teaching hours are fewer. Teaching tasks can only be done at a specific time, there is no extra time to interact with students. And students' spoken language cannot be taught, can only be improved by accepting large amounts of comprehensible input in order to obtain a natural speaking ability. The actual teaching task because of limitations of time cannot provide a lot of comprehensible input, but the WeChat as a spare time social software can solve this problem.

Language learning also requires the creation of a specific acquisition environment, but classroom students only by listening to audio through multimedia, cannot feel the real scene. Through group chat, file sharing, send a small video, WeChat can analog real language environment, to achieve the immersive effect [8, 9].

In summary, WeChat can set up a learning bridge between the classroom learning courses and spare time learning, make up deficiencies of classroom teaching, as an extension of learning beyond the classroom tool is feasible.

5 Empirical Research

This paper selected 30 students (2014) participating in the same computer English Course in Harbin Institute of Technology as experimental subjects. 22 boys, 8 girls. According to the different levels of English, they were divided into three groups, each group 10 students, experimental study period is 18 weeks.

5.1 Research Methods

This paper uses a questionnaire survey to collect data for analysis. Firstly, 30 students must become good friends in WeChat to ensure the circle of friends status be visible to each other. Secondly, Teacher organize the establishment of the discussing group, the teacher's colonization can guarantee the topic not stray from the point, teachers can also participate in the discussion and timely guidance, Thirdly, the group nickname should be changed into real name, so that teachers can calculate the statistical results.

5.2 Experimental Results

5.2.1 Vocabulary Training Memory Module

Teacher assigned group of students to collect words in reference books, literature, brochures and international conferences or the vocabulary required to master, each student can collect 5, within the group cannot be repeated. Teachers organize the discussion in order to recite the words such as humorous mnemonics (split the word, the meaning of each series can help to remember the word, such as "business" can be split into "bus + in + e + ss", through letter pictographic memory, letters "e" and "s" are thought of as "goose" and "snake", the series together is "a goose and two snakes in a bus to do business." According to methods discussed in the group, students share their ideas on reciting collected five words with their friends circle (Fig. 1).

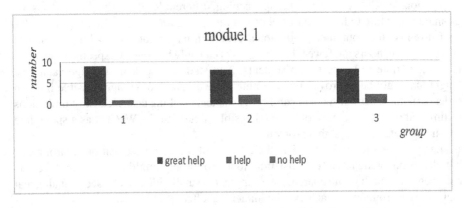

Fig. 1. Vocabulary training memory module. (Color figure online)

5.2.2 Complex Long Sentences and Grammar Training Module

Computer professional English is characterized by long and complex sentences, and sometimes there are several verbs in a sentence, which requires students to have the ability to analyze sentence elements. Teachers show how to analyze the long sentences in each group, the discussion will be related to the grammar, which requires students to explain the relevant syntax, create a "task-based" discussion atmosphere. After discussion, the teacher assigns a task analysis of long sentences to each student, requires to share the results of analyze to a circle of friends (Fig. 2).

5.2.3 Oral Trial Training Module

Based on WeChat, an English software - English Video, is a game in the form of mission passing, can help the students practice speaking, there is no face to face embarrassment, just repeat it after practicing, the game software includes a variety of occasions, background, ritual practice spoken English, practice spoken English and understand cross-cultural etiquette at the same time, students can choose any practice part according to personal preferences, the students can share their progress to the circle of friends, stimulate interest in practice speaking and encourage each other (Fig. 3).

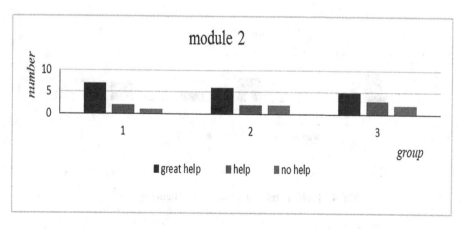

Fig. 2. Complex long sentences and grammar training module. (Color figure online)

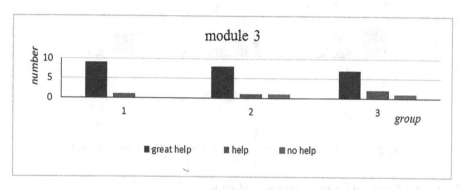

Fig. 3. Oral trial training module. (Color figure online)

5.2.4 Problem Relay Module

Teacher send voice (a short speech content in the professional or the relevant paragraphs in English), translated by the students, who answers quickly, who can get qualifications to set next question, and so on, the students who set a better problem will get extra bonus points (Fig. 4).

5.2.5 The Scenario Simulation Performance Module

Weekly, teacher issues a task performance in English in the group, the contents may be recommendation of a new computer-related products, also can be a simulated international conference scene or even debates. Innovative performance groups can get the actual classroom points, share the performance of video to the circle of friends (Fig. 5).

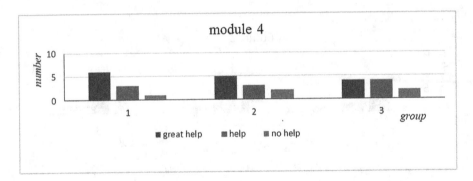

Fig. 4. Problem relay module. (Color figure online)

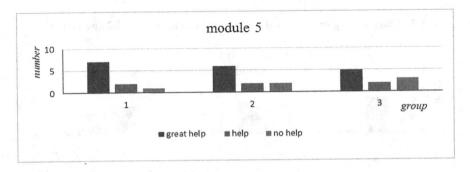

Fig. 5. Scenario simulation performance module. (Color figure online)

5.2.6   Professional English Practical Module

WeChat can share files, concern about the function of public numbers and subscribe to the article, students can always contact any relevant articles to the group, so that the students can discuss learning, related articles may be a new product introduction, or the excerpts of a scientific articles (Fig. 6).

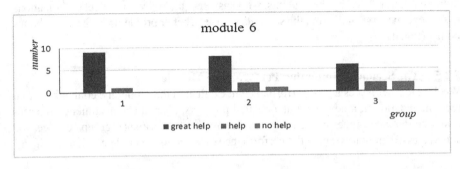

Fig. 6. Professional English practical module. (Color figure online)

5.3   Experimental Results Analysis and Discussion

Chart data analysis showed that the vocabulary memory training modules help students the most, mainly because students in the group discussion, inspired inspiration and motivation of reciting words, words reciting is no longer boring, students are happy to share learning method through WeChat.

For the oral audition module, listening and speaking skills training can greatly increase the opportunities for students to speak English through WeChat and it can improve students' listening and speaking learning enthusiasm. There are training, review, scenario session, simple and practical content, and constantly updated,so that they have courage and opportunity to speak English every day. Students said that they recorded over and over again and practice unconsciously in order to get a high score and exchange within the group, they find that English practice can be so addictive the first time. Students speak positively of English and their enthusiasm is better than before, and the effect is getting better every day.

For the English practical module, students share learning resources and course information via WeChat platform, they can earn first-hand information anytime, anywhere, they are able to develop a vision, deepen curricular learning in their spare time.

For long sentences module part, students said they felt powerless, mainly due to their limited reserves of words, lack of knowledge of grammar. These are the obstacles to analyze long sentences, this kind of task require long-term accumulation in order to enhance the analytical capabilities.

For scenarios show module, as video length restrictions, the students cannot reach their full capacity of oral expression and practical application of English, students can think of limited simulated scenarios, their spare time is unity within the organization, it is difficult for group members to organize simulation performance.

For the audition Relay module, students expressed there is little help for their own, mainly due to the fact that they couldn't fully understand the group's voice, and did not dare to answer, and the students who answered the question will take a long time to send the next voice to the group, the time interval is longer, they gradually lost interest in it.

6   Conclusion

The empirical results show that WeChat-based computer English teaching use a lot of "comprehensible input" ability to produce language output, increasing the opportunities for students to hear and translate in English. WeChat platform sends the bearer resources and transfers teaching resources, expands the curricular learning, and stimulates student interest in learning.

Extensive use of WeChat access to learning resources and participate in learning activities, offers the possibility to improve the proficiency of the language wherever and whenever. It has spread the use of social software for learning computer professional English, meets the needs of contemporary college students mobile learning. It make use of students' spare time to improve English listening and speaking ability and effective way of translation.

The empirical study yielded positive results. But how to establish a scientific and effective WeChat auxiliary computers heard evaluate and test systems and Translating Teaching? How to design a better learning module supervision and guidance of listening and speaking teaching? These questions have to be done in-depth study and practice.

References

1. Zhang, J.: On translation of computer English. J. Changchun Norm. Univ. (Nat. Sci.) **28**(1), 121–123 (2009)
2. Xu, X.: Interpreting metaphorical statements. J. Pragmatics **6**, 1622–1632 (2010)
3. Wang, J.: The research and practice of bilingual teaching methods in computer English. Comput. Educ. **4**, 80–82 (2010)
4. Jiang, X.H.: Research on pedagogic approach of specialized English for computer science. J. Luoyang Norm. Univ. **31**(3), 110–112 (2012)
5. Miao, N.: English mobile learning strategies in universities and colleges based on WeChat. Chin. Acad. J. Electron. Publishing House, 136–140 (2016)
6. Yan, K.H.: The discussion and research of computer English teaching methods. Overseas Engl. **350**, 102–103 (2012)
7. Feng, C.L.: An overview on Krashen's input hypothesis of second language acquisition. J. Hubei TV Univ. **30**(8), 99–100 (2010)
8. Yao, W.H.: Research on teaching of computer English. Comput. Educ. **13**, 88–91 (2011)
9. Liu, W.: Optimization of teaching college English listening and speaking on modern educational technology. Chin. Acad. J. Electron. Publishing House **10**, 116–119 (2013)

Exploration of C Language Practical Teaching Method Based on Project Learning

Huipeng Chen, Yingtao Zhang[✉], and Songbo Liu

Harbin Institute of Technology, Harbin 150001, China
yingtao@hit.edu.cn

Abstract. This paper proposes a practical teaching approach in C programing language based on project learning. Having performing experimental teaching in class, this approach obtains pretty good teaching results, although it presents some flaws. This paper briefs the process of teaching, analyses the problems which are found during teaching and provides the solutions accordingly.

Keywords: Project learning · C language · Practical teaching

1   Introduction

As a component of computer languages, C programing language is widely used for computer language teaching in most of higher colleges and universities [1–3]. With the rapid development of computer languages, many languages, such as C++, Java, C#, play an important role in computer field [4]. C language has long been kept in computer language teaching due to its distinct features. With the development of inserting system, C language has its usage area expanded.

This paper offers brief analysis in terms of problems arisen from C language teaching. During C language teaching, a large number of books (including classical books) put more focus on rules of C language, which has a huge impact on teaching [5]. However, due to the distinct characteristics of C language, students who usually command the rules have difficulty in programing. Noting this point, many teachers add a plenty of experiments to improve students' ability of programing. Although performing lots of experiments, students still are confused when they are confronted with a specific case. And this is currently criticized by many enterprises which believe that university education do not meet the work need. Then where is the problem? On the basis of lots of practical teaching, this paper offers several problems: (1) C language students have learned is just a kind of language; (2) Although relevant curriculum involves the content of modelling during programing process, these contents are isolated; (3) The experiments usually contain small and individual programs rather than a complete system, which makes students unfamiliar with the concept of system. How these problems will be solved in teaching? This paper explores a teaching method in an optional class, which expects to tackle the problems above.

© Springer Science+Business Media Singapore 2016
W. Che et al. (Eds.): ICYCSEE 2016, Part II, CCIS 624, pp. 21–25, 2016.
DOI: 10.1007/978-981-10-2098-8_3

2   Constructing a Project

The primary task is to design a project which needs to be achieved during the course teaching. Then we need to analyze the knowledge involved in the project, and gradually interpret the knowledge according to the proceeding of the project. After the target in one stage is set down, the questions will remain for students to deal with. At last, all the knowledge is integrated into a complete system, thereby achieving the aim of the project [6].

To stimulate the interest, the projects in this course are all game programming. Game programming involves lots of knowledge, such as computer graphics, artificial intelligence, interrupt, keyboard operations, file reading and mathematical modeling. A small game programming is a complete system and students are interested in game programming. Tetris, for example, is a game familiar to all of us. To carry out a complete Tetris game, students should get a understanding of graphic modeling and displaying, movement and overturn, controlling of falling speed and so on [7, 8]. The task aims at a demonstrable program in a course period.

3   Analyzing and Interpreting the Project Knowledge

As Tetris program contains an amount of knowledge, students who have not designed large programs may be confused with Tetris and they do not know how to set out to do it. Teachers' task is to analyze the techniques needed in the project and gradually interpret them.

(1) The first problem need to be solved is graphic display. Teachers should clearly tell students about graph patterns and text patterns in DOS setting, transformation between different patterns and how to carry out them in C language. For avoiding complex operating system and compiling environment, this course applies DOS operating system and Borland C3.1 compiling environment.

(2) How to show graphs is the next step. Teachers give an example of how to draw a circle or a few lines in a projection screen. By drawing a clock for instance, a teacher can not only clearly show how to display graphs, but also explain coordinates in two dimensions.

(3) The next step is about graphical movement. The pointer's moving can clearly illustrate the way of graphical movement.

(4) The sequent stage is to solve time correspondence. Teachers should explain the working principle of system clock and how to obtain the system time in C language. When questions above are dealt with, students can write a grogram which simulates a simple clock on their own.

(5) Tetris square falling is automatic and beyond the control of the operator, but the falling speed can be controlled by keyboard. To carry out this function, we should explain the interrupt principle in DOS, and then query the interrupt vector number $(0 \times 1C)$ associated with the time. How to design a interrupt program and how to put the entrance address into interrupt vector table is complex to explain. And the interrupt service program's written and operation is also a complicated knowledge

point. Many students need to spend much time to understand them if they do not learn the course of computer principles.

(6) When integrating the interrupt and graphical techniques, students can write a program which controls Tetris square' automatic falling using the clock. Up to now, students have integrated different knowledge. This process involves introduction of knowledge, explanation of programming techniques and training by writing game programs, which must be interesting for students.

(7) How to model Tetris squares is the most complicated issue. There are many methods to solve it, and teachers should motivate students to explore models on Tetris squares. Teachers had better provide one or two ideas to guide students to model.

Tetris squares have seven shape types, shown in Fig. 1.

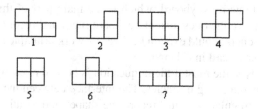

Fig. 1. Seven basic shape types of Tetris squares.

Idea 1: seven graphs can be drawn in computer using a drawing tool and a graph file is created to store in hard disk directory, which needs to understand the file operation and the format of graph files (BMP, JPG, TIFF and so on). What needs to do is opening and decoding graph files in programming software and displaying graphs onto the screen.

Idea 2: Every Tetris square graph comprises four small squares, but the difference is the locations of four squares leading to distinct external features. Using digits to represent these squares is one idea of modelling Tetris squares. We can construct a 4×4 data structure shown in Fig. 2. This kind of data structure can represent different shapes of Tetris squares by showing different locations of squares. Digit 1 means the corresponding location has a square and digit 0 means the opposite. In this way, we can construct a $7 \times 4 \times 4$ three-dimension array which represents seven different kinds of Tetris. For describing four kinds of rotations, a $28 \times 4 \times 4$ data structure is needed, which can express basis squares and also rotated squares. As the array contains only two digits 0 and 1, could we consider the binary code? Could the model in Fig. 2 be represented as $0 \times C600$? If the answer is yes, a hexadecimal digit can represent a kind of Tetris square. It has no problem in theory, but how should we construct the relationship between digits and graphs and the relationship between locations and bits? These ideas aim to implement programs and explain modeling methods. It is worth noting that this method is not the best. A problem will arise from programming, because the Tetris square comprises four squares following given locations rather than a complete square.

Fig. 2. Data structure.

It is a kind of teaching thought that teachers do not provide ultimate correct answer but provide basic ideas which students can follow to design better solutions. We can say that for practical projects, there is no perfect solution, but only better ideas.

(8)   The next question is how to make players to control Tetris squares' left-right movements and rotations by keyboard, which is the main part of this program. What needs to do is solve the questions of keyboard input and recognition of keyboard scan codes. Teachers should explain keyboard's work principle and corresponding techniques for keyboard in C language.

(9)   The last and maybe the most difficult question is how to make Tetris squares stop falling when encountering obstacles. The interface between the Tetris square and the obstacle is irregular, we must record the shapes and locations of fallen Tetris squares. As previous design does not consider recording the fallen locations, students find the code disordered and need to make up for it. Under this situation, teachers need to propose systematic ways of thinking.

Systematic thinking depends on integral need consideration which decomposes the overall system into several parts and integrates parts into one system. If the two questions are solved, the program can be carried out smoothly, otherwise, one problem is likely to damage the entire system, thereby wasting time and reducing efficiency. If it happens, this accident could make a deep impression on students who can learn from the lesson that systematic thinking is indispensable for a practical project.

4   Implementing the Project

During the implement process, students need to write complete systematic programs and teachers just explain key points. The most difficult part is how to debug, as lots of problems can arise during programming process. The problem can be solved by tracking by step, but typically one problem cannot be located correctly. How to design debug methods is a problem which students encounter. Teachers need to tell students by which way students can find the problem.

There are two reasons why this course chooses games as programming cases: one reason is that game can stimulate students' interest in programming; the other reason is that game interface is usually in graph form which can clearly show the problems, so students can easily find the problems and debug them.

5   Conclusions

By three years' undergraduate teaching experience, we find several advantages of this course: (1) It motivates students' interest in programs. (2) It improves students' systematic programming competence. (3) It improves students' debugging skills. (4) It promotes students' confidence on programming. But we also find several problems: (1) Students who persist in taking class can carry out the project, while absent students cannot due to the lack of the corresponding knowledge. (2) As there is lots of new knowledge and they should be understood and enforced immediately, it is a challenge for students who have a slow understanding, so many students choose to give up. (3) Up to now, involved projects are Tetris and Gobang.

This teaching method exploration applies project based teaching, achieves teaching goal by setting the project goal, analyzing project techniques, explaining knowledge and implementing programs. We believe programming teaching such as C language is a kind of practical teachings, and skills need be acquired by training and practicing, so we abandon traditional teaching method, experience practical teaching based on the project, which obtains better effect.

References

1. Yang, J., Xu, D., Liu, H., Qi, W.: Research on the Teaching Reform of C Language Course. Science Education Article Collects (2014)
2. Li, N., Wang, W.-H., Wang, Y.: Teaching innovation beyond "Path-depended" on C language course. J. Anyang Inst. Technol. **2**, 033 (2012)
3. Shi, H., Zhang, J. M.: Discusses several difficulties' teaching method in C language course. Comput. Knowl. Technol. (2012)
4. Wu, L.-J., Shen, H., Zhang, H.-H.: Exploring and practice of C language course designing teaching model. J. Shenyang Normal University (Nat. Sci. Edn.) **1**, 029 (2012)
5. Jin-fang, T.: Exploration on teaching method of C language course. Higher Educa. Forum, 030 (2007)
6. Kay, J., et al.: Problem-based learning for foundation computer science courses. Comput. Sci. Educ. **10**, 109–128 (2000)
7. Lee, T.-S., Yi, S.-H.: The effects of tetris game on the cognitive abilities of students with mental retardation. J. Korea Game Soc. **12**, 77–85 (2012)
8. Chen, X., Lin, C.: Tetris game system design based on AT89S52 single chip microcomputer. In: Third International Symposium on Intelligent Information Technology Application, IITA 2009, pp. 256–259. IEEE (2009)

Exploration of Integrated Course Project Mode of the Internet of Things Engineering Based on the Relevance Theory

Qiaohua Feng[✉], Wenjie Zhao, Xiaoyu Yu, and Yunbo Shi

Institute of Measurement-Control Technology and Communications Engineering,
Harbin University of Science and Technology, Harbin, 150080, China
{fengqiaohua80,shiyunbo}@126.com,
{zwjsky888,13904808427}@163.com

Abstract. Based on analysis of the present situation and problems of course project of the Internet of Things engineering, we propose a relational integrated course project pattern around three levels of perception layer, network layer, application layer of the Internet of Things system. The realization of a complete Internet of Things system is divided into three course projects to complete three key points, which may eventually make a complete system of things. Through the link among the three integrated course projects, knowledge of four years will be connected together and form an organic whole. We use a team performance and examination methods of the process-oriented examination, project paper and oral examination for the integrated course project in order to improve students' cooperation ability, expression ability, communication ability and other integrated quality.

Keywords: The Internet of Things engineering · Relevance · Integrated course project · Process inspection

1 Introduction

Internet of Things (IoT) as a critical information infrastructure has been used in all areas of society, the demand for high-level IoT talents is also increasing, Internet of Things engineering is set up for the urgent talent demand of IoT construction and application, and employers have higher requirements for engineering practice ability of the graduates. But the supply of college graduates removes from the demand for talents, and the demand for talent of the enterprises is far not satisfied. In order to improve the overall quality and innovative spirit of higher education talent, practical teaching activities are very important [1]. "Excellent engineer education and training program" by Ministry of Education puts forward new demands for the quality of innovative engineering education of undergraduate, and leads engineering university back to the essence of engineering education rationally [2]. Therefore, the important work in the development of the Internet of things engineering is to improve practice teaching system of the Internet of Things engineering, strengthen practice ability and engineering innovation ability of undergraduates and train qualified engineering and technical talents of IoT.

© Springer Science+Business Media Singapore 2016
W. Che et al. (Eds.): ICYCSEE 2016, Part II, CCIS 624, pp. 26–31, 2016.
DOI: 10.1007/978-981-10-2098-8_4

Among them, course project is one of important integrated practical teaching and an important part of teaching. Course project is an important part of training students to apply the knowledge learned to discovery, present, analyze and solve practical problems and exercising the practical ability of undergraduates, and is the specific training and study process of practical work ability [3]; course project focuses on training students to apply basic theory to solve practical engineering problems, improves the ability of students applying their integrated expertise, and occupies an important position and role in the teaching programs of Internet of Things Engineering [4].

Based on the characteristics of Internet of Things Engineering, combined with the actual situation of our university, we explore an integrated course project mode with its own characteristics.

2 Analysis on Practice Teaching Status of Course Project of Internet of Things Engineering

The Internet of Things Engineering is a major with high engineering application, but in the actual teaching of many colleges, due to the limitations of teaching hours and laboratory equipments, practical teaching aspects is very weak, course projects are mostly academic and theoretical contents, and technical, practical and integrated contents continue to weaken, actual operation of undergraduates is less in the real sense, so course project is like "writing a report", and has only theory and no practice. At the same time, course project is independence, relevance between course projects is very less, which cannot allow students to connect the knowledge learned in the four years to form an organic whole, therefore, students have a sense of "what is the use of this course project".

In addition, course project has the single inspection method of the course project report in general, the focus of this method is to inspect the theoretical knowledge mastered by students, to some extent, which focus on book knowledge and ignore the capacity of students guided by key teaching points of teachers; and most of the course project reports are plagiarism seriously, which cannot really inspect the students' practical abilities.

Therefore, based on the characteristics of Internet of Things Engineering, combined with the actual situation of our university, we explore a integrated course project mode with relevance, which allow the students using knowledge learned in many theory courses to design theoretically and implement it in practice during a course project, and the object obtained in the course project will be used for the next course projects to form a system.

3 Construction of Integrated Course Project Mode

3.1 Knowledge System of the Internet of Things Engineering

Because the Internet of things engineering is an emerging discipline, subject system containing knowledge structure and courses has not been set up independently. But its technical architecture has been basically determined from the perspective of technology and industry, composition of IoT system is very long, the architecture of the system can be roughly divided into perception layer, network layer and application layer, and each level involves many subdivision fields.

In order to achieve certification standards of the Internet of Things Engineering, the course system of the IoT is consistent with the industry demand to the IoT technology, three layers IoT technology architecture and the effect of supporting disciplines. Therefore, dumbbell-shaped linear knowledge structure is a professional and reasonable positioning to the Internet of things engineering in our university, it means that undergraduates should grasp the knowledge of perception layer and application layer at the both ends of the IoT system and understand basically the knowledge of the network layer in the middle of the IoT system, that is, students should have an important mastery to the knowledge of data acquisition and control, data area (short) transmission and communication, data access and heterogeneous interconnection, data storage and position, calculation, mining and application of data and so on in the data flow of the IoT system, and have a basic grasp to the knowledge of public network transmission of data, data communication and management and so on. The specific knowledge system is seen in the reference [5].

3.2 Construction of Course Project Mode

According to knowledge system of the Internet of Things Engineering in our university, around a complete IoT system with three levels of perception layer, network layer, application layer, the implementation of the entire system can be divided into three integrated course projects to solve three key points, and around each key point, a complete system can be achieved in senior year from theory to implementation of course project.

Course Project 1 is "Electronic Technology of Analog-to-digital Conversion and Signal Analysis", the key point to address is the conversion of analog signal to a digital signal and processing method and analysis of analog and digital signals; Course Project 2 is "Design of Sense Terminals and Wireless Nodes", the key point to address is identifying and getting information, collecting the physical events and data in the physical world, and realizing the information perception and recognition in the outside world; Course Project 3 is "Mobile M2M Software and Network Design", the key point to address is completing the transmission and processing of sensory information with high reliability and high security. The three course projects are together connected to form a small system of IoT, which completes perception and identification of data in the outside world, transmits the data through the network, and analyzes and processed the data by the receiver.

Course Project 1 is carried out in the fourth semester, the course practice hours are two weeks, and the related theory courses includes Circuit, Electronic Technology, Signals and Systems, Digital Signal Processing. This course project is to train the basic knowledge and basic skills, and requires students to design and weld an analog-digital conversion circuit based on four courses and master the method of signal processing.

Course Project 2 is in fifth semester, the course practice hours are two weeks, and the related theory courses includes Principle and Application of RFID, Internet

Communications Infrastructure, Principle and Application of SCM, Perception and Control Circuit of IoT, Sensor Technologies and Applications. The signal acquired by the sensor or RFID is converted into a standard voltage signal by measurement circuit, and then changed into digital signal through the analog-digital conversion circuit made in course project 1, then connected a microcontroller and wireless terminal to realize a specific function in the C programming language, which finally forms a "sense terminal".

Course Project 3 is in the seventh semester, the course practice hours are also two weeks, and the related theory courses includes M2M Communication Technology and Application, Computer Network, The JAVA Programming Language, Wireless Sensor Network Technology. Through networking protocol some "sense terminals" in course project 2 form a network, and communicate among the terminals, and human-computer interface via JAVA programming language analyzes and processes data.

Above all, the knowledge points of four college years install in series together to form an organic whole by these three integrated course projects, which can improve the students' professionalism and ability to solve engineering problems, and enhance the sense of collaboration. The relationship between course projects and related theory courses is shown in Fig. 1.

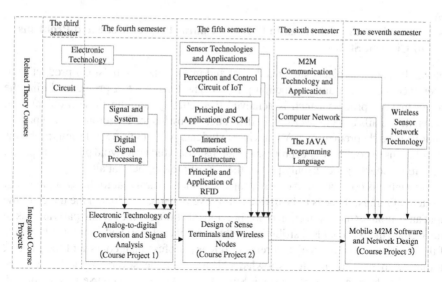

Fig. 1. The relationship between course projects and related theory courses

3.3 Implementation Processes of Course Project Mode

Correlation between the course projects determines the relevance of the construction of related theory courses. The constructions of course projects and theory courses closely link with teachers engaged in teaching these courses, so the strength and coordination of the teachers is the key. Students are not only beneficiaries but also inspectors. Therefore, the relational integrated course projects are inseparable from the active participation of students.

In order to ensure the implementation of an integrated course project and cultivate students to solve practical engineering problems and improve innovative ability, set the students to center, and the whole course project is all under the guidance of teachers in order to help students to answer questions.

During a course project, three students make a group, and a group has a topic different from others, which requires members solidarity and cooperation and a reasonable division of labor, this approach enhances the sense of collaboration and urges each other to learn together.

In addition, we introduce the process inspection during the integrated course project. In addition to attendance and working attitude, project report and oral examination, process inspection is important during a course project and more than 60 % score. Instructor provides the task of a course project, and the task will be divided into several stages such as the overall design, circuit design and simulation, circuit welding and debugging, experimental data processing and other stages, each stage accounts for the score. The students are graded by the concrete performance and task achievement at each stage, which can urge students to speed up the design and ensure to complete the task.

3.4   Analysis of the Relationship Between Course Project Mode and Problems of Current Status

During the course projects students design and make a hands-on "small product" and experience the design and production process. There are a lot of problems during the project making process which is achieved theoretically, solving these problems can deepen the details easy to overlook in theoretical knowledge. Therefore, course project is not just the theoretical one. Meanwhile, Students write their own design, problems and solutions of course project process in their papers, which can avoid plagiarism paper. In addition, course paper is only part of the course project, but not all.

Through this correlation of course projects, students learn that for the completion of a system of things, it needs to be divided into several parts, considered the match of the input and output at the connection of the previous section and the second parts and taken into account the reasonable distribution of the error of each part and a reasonable choice of the devices and so on, which is related to many professional courses of the Internet of things, so that fragmented knowledge learned in the several semesters can be connected to form a complete system. Students can know each course is not isolated, there is a connection between courses, where can use each course. These can lay the foundation for graduation project of senior graduation and provide protection for future employment or student to continue their studies.

4   Actual Results of the Course Project Mode

From the experiences of course project papers, we can see the students did not know how to start at the beginning of the course project, then resolved the difficulties step by step under the joint efforts of group members, and finally produced a small product successfully, even if some final products did not achieve the desired targets, but also they learned a lot. These indicate our course project mode can improve students' hands-on skills. And the quality of paper is also greatly improved.

We learned from the interviews of postgraduate candidates exempt from admission exam for over two years, the most impressive course is our course project, they still remembered the content and process of course project, and heard from the feedback of graduate students, they took part in their subject more easily during the post-graduate. These indicated that this course project mode can really deepen students' understanding and mastery of knowledge.

The teaching mode takes more time and energy of teachers and students, but students really experience the "putting them into practical use".

5   Conclusion

Based on the knowledge system of Internet of Things Engineering in out university, we construct an integrated course project mode with relevance, it effectively connects the knowledge points of college to form an organic whole, and introduce process inspection to motivate students to learn actively, improve students of professionalism and ability to solve engineering problems, and enhance the sense of collaboration. There is also reference function for other relevant majors.

Acknowledgments. This research was financially supported by Higher education teaching reform project of heilongjiang province in 2014 (No. JG2014010794) and Higher education teaching reform project of heilongjiang province in 2013 (No. JG2013010298).

References

1. Ju, G., Zhou, G., Chen, R.: Practice teaching system of based on excellent engineer training. Exp. Sci. Technol. **13**, 26–29 (2015)
2. Teng, W., Liu, Z., Fang, J.: Study of cultivation mechanism of university-enterprise cooperation innovation talents for Internet of Things Engineering facing the "excellence initiative". EDUCATION GUIDE, pp. 37–38, 40 (2015)
3. Cao, L., Li, Y., Zhang, A.: Exploration of researchful program design based on excellent engineer training. China Electric Power Education, pp. 36–37 (2015)
4. Cai, C.: The Construction of Practice Teaching System of Internet of Things Engineering based on CDIO. Education, pp. 80–82 (2015)
5. Shi, Y., Liu, B., Yang, M., Han, T.: The Thinking of Multidisciplinary Integration to Build the Internet of Things Engineering Specialty Course System. Internet of Things Technologies, pp. 108–110, 113 (2016)

Exploration on the Application of Microlecture in Presentation-Assimilation-Discussion Class

Zhifang Wang$^{(\boxtimes)}$, Jinjin Dong, Bing Zhao, and Jiaqi Zhen

Department of Lectronic Engineering, Heilongjiang University,
Harbin, Heilongjiang, China
zhifang.w@gmail.com

Abstract. As the rapid development of information technology represented by mobile internet, internet of things, social computing, society has entered the age of big data, information technology and education teaching are also merged gradually, microlecture as a new application of education and teaching took the wide attention of scholars. This paper analyzes the disadvantages of PAD class and takes microlecture to apply to two stage of PAD class.

Keywords: Social computing · Microlecture · PAD class

1 The Appearance and Significance of Microlecture

The rapid development of information technology, especially the Internet of things, cloud computing, social networking, social media and the progress of information technology, data is growing and accumulating rapidly with the great speed, the time of big data has come [1]. As a data-intensive science social computing has had a huge impact on the breadth, depth and scale of the collection and analysis of data, in this case the microlecture was born quietly.

The rise of micro class has its roots. On the one hand, in the past twenty years, the development of educational informationization, the resources of course video not only can't conform to the age of the Internet users' attention model, but also hard to meet the actual demand of a gleam of teachers and students, and the bandwidth and internet speed caused the waste in the process of education resources construction and application. On the other hand, under the condition of the era of big data, with the increase of network bandwidth, the development of video compression and transmission technology, the rising popularity of wireless network, and the increasingly popular mobile terminal, the Internet entered the micro age. Weibo, WeChat, micro film blustery, rapidly in popularity at the rate of more than people imagined. Under such a background, the word microlecture arises at the historic moment, and quickly be widely adopted by online educators [2].

The emergence, design, development and application of microlecture have very important strategic significance and realistic value for school education and social education which in the age of mobile learning. In terms of school education, the microlecture not only becomes an important education resources

© Springer Science+Business Media Singapore 2016
W. Che et al. (Eds.): ICYCSEE 2016, Part II, CCIS 624, pp. 32–35, 2016.
DOI: 10.1007/978-981-10-2098-8_5

for teachers and students, but also formes the basis of the school educational teaching mode reform. It has important practical significance for both students' learning, teachers' teaching practice and teachers' professional development. Not only that, the development of micro class will also trigger a new round of basic education in the development of the digital teaching reform [3].

In recent twenty years, more and more schools start trying online education, students study outside of the classroom on summer and winter vacations with the aid of network has become a beautiful scenery. And in the practice of online learning, online video features shortmicrolecture, has become an important learning resources. At the same time, it is often necessary for the teacher's classroom teaching practice to use some short teaching video for the explanation and demonstration in the process of classroom teaching.

In the last two years, with the rise of microlecture, a lot of online education companies also began to try the commercial application of the microlecture, especially online tutors, online continuing education geared to the needs of specific people, and share skills to public [4]. These online education companies, try to share the relevant knowledge and skills in the form of miniature teaching video by making use of the combination of online and offline, or simply by means of the online mode to build an online learning environment of sharing knowledge and skills.

2   PAD Class

2.1   The Concept of PAD Class

Presentation-Assimilation-Discussion (PAD) class is a new mode of classroom teaching reform, which be proposed by Zhang Xuexin who is psychology department of Fudan University. Through to mobilize and exert students' autonomous learning, reduce the burden of teacher's teaching, improve the teaching quality, enhance teaching effectiveness. The application of psychology of wisdom in the education teaching practice.

The core idea of Sub-class is to put the half of the class time allocation for teaching and the other half assigned to the students to discuss for interactive learning [5]. Compared with the traditional classroom, Sub-class emphasizes on teaching before, teachers teach students in first. Similar discussion-based classroom, PAD class emphasis on students to interact with students, teachers and students, encourage independent learning. The key of Sub-class innovation lies in the lectures and discussion time stagger, let the students have a week after class time independent arrangement study, undertake personalized internalized absorption. In addition, in the assessment method, Sub-class emphasis on the process of evaluation, and pay attention to different learning needs, so that students can determine the input of the curriculum according to their individual learning goals. Sub-class teaching is divided into three processes in time, respectively, Presentation, Assimilation, Discussion, so PAD class can also be referred to as PAD class [6].

2.2   The Shortcomings of PAD Class

In PAD class teaching mode, import link of teaching knowledge teaching, aims to enable students to understand theme teaching to form the basic framework of knowledge system. Therefore, students need to internalize, and to grasp learning the difficult point after class. But students do not have access to a large amount of information in a short time, as well as master and memory a large number of concepts, principles, formulas. The lack of this series of knowledge will lead to a lot of problems in the process of internalization, then the effect of student learning will be greatly reduced.

For students, although classroom participation helps students to turn knowledge into an organic part of their knowledge system. And classroom discussion is the most effective way to participate in the classroom [7], but because the students in the learning process to absorb the degree of knowledge is different, each student's learning experience, experience and the question is also different. So students in discussions related to the theme of the content will be more broad, lead team members in the mutual exchange is not easy to reach a consensus, it is not conducive to the second level of the whole class discussion, not conducive to the realization of common progress between the students.

3   The Application of Microlecture in PAD Class

Teachers can take advantage of microlecture to sum up the lessons, students can be more profound understanding of knowledge content and clues, and thus targeted for study [8]. According to the their mastery degree of knowledge to find more development content, which can effectively understand the textbook knowledge, break through learning difficulties, to achieve targeted, enhance learning effect. Meanwhile, teachers can set up seminars by using the micro-class to explain the learning process typical problems and help students organize learning.

Students can use the microlecture to consolidate the content of the teacher lectures, a better understanding of textbook knowledge, a breakthrough learning difficulties, thus targeted learning content and expand learning. And students can be prepared to discuss the content of the class effectively, discussion topics will be even more clear, students can efficiently reach group consensus, then smoothly for whole-class discussions at the second level, to achieve common progress between students.

For the lesson of mathematical foundations in information security, we tempt to induce the microlecture into the PAD class. In classical class, the teacher one-time teaches knowledge within limited class time. When the students absorb knowledge after class, the textbook is only the reference materials. In this condition, the learning effect may be affected. If the teacher can summarize the obvious knowledge point to make the corresponding microlecture, the students have better access to absorb after class. In the lesson of mathematical foundations, we initially divide into four modules: number theory, modern algebra, combinatorial mathematics and chaos. For each modules, the key point and the corresponding question types are summarized in documents. But we find that silent documents can't guide the students to learn, can only provide learning materials. Then we further detailing the knowledge and make the microlecture for each key point. For example, modern algebra can be divided into three points: group, ring and field. For group, we further provide three concepts: symmetric group, cyclical group, ideal. For practical teaching, we explain the concept of group in detail and present the new point simply. The detail of symmetric group, cyclical group, ideal require the students to accept and absorb. In this way, the students understand the knowledge by self-study deeply. In addition, the student can explicit the study objection and adopt the new content quickly by reviewing the knowledge the teacher explained in class. Comparing with the traditional microlecture, this course is relatively more system. Accompanying with the school curriculum and teacher guidance, the students will actively learn the course of knowledge points, and the microlecture utilization rate is higher.

4   Conclusion

Based on the background of the era of large data, this article introduces the appearance and significance of microlecture in detail, introduces the application of microlecture today, and analyzes PAD class teaching mode in detail, then expounds the concept and significance of PAD class, and analyzes the two shortcomings of PAD class. Based on this, this article come up with an idea that make microlecture apply to PAD class, so that it can largely made up for the existing inadequacy of PAD class mode, and it can make the application of microlecture become a useful complement to PAD class.

References

1. Meng, X., Li, Y., Zhu, J.: Social computing: the opportunities and challenges of the age of big data. J. Comput. Res. Dev. **50**(12), 2483–2491 (2013). (in Chinese)
2. Guan, Z.: Microlecture. J. Chin. Inf. Educ. Technol. **17**, 14 (2011). (in Chinese)
3. Educause. 7things You Should Know About Microlecture [EB/OL] 12 April 2013.   http://www.educause.edu/library/resources/7-things-you-should-know-about-microlectures
4. Tiesheng, H.: Microlecture: the new trend of the development of education information resources. Electrified Educ. Res. **10**, 61–65 (2011). (in Chinese)
5. The sub-class of Fudan university. Discuss the reform. [EB/OL], 20 March 2015. http://www.duifen.org/ (in Chinese)
6. Li, H., Yunyong, W.: The analysis of interactive teaching mode. J. Shenyang Normal Univ. Soc. Sci. Ed. **31**(3), 77–79 (2008). (in Chinese)
7. Zhen, H.: The comparative study of classroom teaching in China and the United States University. High. Educ. Dev. Eval. (2), 106–111 (2007). (in Chinese)
8. Yu, S., Chen, M.: The design of micro lesson based on unit learning platform. Open Educ. Res. (2), 100–110 (2014). (in Chinese)

Research and Practice on College Students' Innovation and Entrepreneurship Education

Hui Gao[1], Zhaowen Qiu[2(✉)], Zhengyu Liu[1], Lei Huang[1], and Ying San[3]

[1] Harbin Huade University, Harbin, China
44117252@qq.com
[2] Institute of Information and Computer Engineering, Northeast Forestry University,
Harbin, China
qiuzw@nefu.edu.cn, 85312371@qq.com
[3] Harbin Guangsha University, Harbin, China
249590520@qq.com

Abstract. Strengthening the education of innovation and entrepreneurship is one of the important tasks of China's higher education reform and development. Entrepreneurship Education should focus on setting pioneering genetic code for future generations. Essentially, it is an education innovation-oriented entrepreneurial revolution of human resource development. Technological innovation is strategic support to improve social productivity and comprehensive national strength, which should be placed at the core of national overall development. For China's higher education, it proposes more new requirements. Serious discussion on the innovation education of college students is needed. Through practice, improvements in the quality of innovation and entrepreneurship education will be achieved.

Keywords: Innovation and entrepreneurship · Education mode · Talent cultivation

1 Innovation-Driven Development Strategy Puts Forward New Request on Talent Cultivation in Colleges and Universities

Report of the 18th CPC national congress pointed out that Technological Innovation is strategic support to improve social productivity and comprehensive national strength. It must be placed at the core of national overall development. Innovation-driven development strategy is put forward under the background of economic growth pattern transition. Since China's reform and opening up, the economic and social development is going up by leaps and bounds. Now, labor costs are rising, advantages of labor-intensive industries are no longer evident. Since the overdependence on resources exploitation, open development has a serious impact on environment bearing capacity. The rich labor resource which propped up the traditional economic development pattern is constantly losing its edge, and can hardly play an major role. But, growth based on knowledge innovation and specialized human resources can not only form internal increment of capital gains, but also make the traditional productive elements generate incremental

© Springer Science+Business Media Singapore 2016
W. Che et al. (Eds.): ICYCSEE 2016, Part II, CCIS 624, pp. 36–44, 2016.
DOI: 10.1007/978-981-10-2098-8_6

benefits. And with increasing size of the entire economy benefit, the limits to growth in the traditional sense will be broken through. So, China proposed implementation of the innovation – driven development strategies timely to accelerate the conversion from low cost advantages to innovation advantages, and to search for new power sources for sustainable development. It has important practical significance in enhancing the quality and efficiency of economic growth and accelerating the transformation of economic development.

Innovation-driven development strategy definitely places science and technology innovation at the core of the national overall development. This shows that instead of the traditional labor and resources exploitation, China's future development will rely on science and technology innovation. Development model transformation has brought the need for innovative talents. It proposed new and higher requirements for China's higher education.

First, in respect of personnel training, colleges and universities should study further on the theories which support for innovation – driven development strategies. According to research, the theory originated from Schumpeter's theory of economic development and the Romer's new economic growth theory that play a decisive role in economy. Innovation – driven development strategy is the innovative application of the theory. An in-depth study of these theories is the foundation of innovation and entrepreneurship education.

Secondly, it should be fully understood that the essence of innovation – driven development strategy is to develop innovative economy. Implementing innovation – driven strategy and developing innovative economy are largely dependent on knowledge and technology system innovation and development. Innovative economy is the embodiment of resource conservation and environmentally friendly requirements, which is based on knowledge and personnel, driven by innovation, and marked by economic innovation industry. The only implementing subject of innovative economy is personnel with innovative ability, this requests us to start from the essence, grasp the essential basic knowledge innovation.

Finally, it should be made clear that implementing the innovation-driven development strategy demands new labor resource. Large amount of innovative personnel of high-quality are needed to the implementation of the development strategy. Training and proving innovative personnel becomes the urgent task for our country's higher education.

In short, we should lay the foundation on a variety of theories, follow the guide of innovation-driven development strategy, and be full engaged in innovation and entrepreneurship education. It requires colleges and universities to seriously study and build innovative personnel education mode.

2   Problems Existing in College Students' Innovation and Entrepreneurship Education

Some universities have already began to explore and practice in building the training mode of the ability of innovation and entrepreneurship. In the development of innovative entrepreneurship training there still exist some problems that have a direct impact on the effect of cultivating students innovation ability.

2.1 Insufficient Knowledge of Cultivation of Innovation Ability

Some universities generally do not attach importance to training of creative ability and lack of awareness about the depth of the cultivation of creative ability, which failed to establish a correct concept of innovation and entrepreneurship abilities.

2.2 The Setting of the Course System Remains to Be Improved

The creative ability training is still carried out in extracurricular link in some colleges and universities, which is non-standard. In the teaching plan and credit system, only one or two employment-oriented subjects are carried out such as Employment Entrepreneurship Guidance Of Employment or Basic Entrepreneurship, lacking the combination of theories of innovation ability cultivation and practicing activities.

2.3 The Teachers' Level of Creative Ability Training Remains to Be Improved

For the creative entrepreneurial teachers, we lack the guiding teachers with innovation entrepreneurship theories and entrepreneurship practice experience. Many colleges and universities make great efforts to vigorously introduce teachers who are highly educated and teaching-research talents with high diplomas and titles. They have certain academic ability but innovation ability. They have certain professional theoretical knowledge but industry development ability.

2.4 Classroom Teaching Remains to Be Optimized

Teaching contents are confined to the operating level and skill level, and the cultivation of students' innovative consciousness is ignored. The cultivation of students' innovative entrepreneurial spirit is not enough in the teaching methods aspect, so that the subjectivity and enthusiasm of students can not be given priority during their passively accepting knowledge. The assessing methods are not varied, which does not include the assessment of the student's innovation ability evaluation, lacking the strong innovative entrepreneurial atmosphere.

3   The Exploration of Creative Education Mode

Innovation entrepreneurship education mode refers to a relatively stable strategy system of entrepreneurship education programs and methods in a certain environment innovation under the guidance of the entrepreneurial ideas. The university students' innovation entrepreneurship education pattern will be preliminarily explored, based on the experience of the innovation entrepreneurship education in colleges and universities at home and abroad.

3.1   The Establishment of the Concept of Education

During the era with Knowledge economy as the leadership of the new economy, new requirements will be put forward for our country's higher education. Students are supposed to have not only a lot of theoretical knowledge but also a scientific outlook on life and social responsibility, and more importantly they should have strong practice ability and innovation ability. Innovation concept of entrepreneurship education should give the capacity and quality of college students' innovative entrepreneurship a high priority.

Developing the innovation ability can not happen overnight, but a long systematic process. So we need to innovate ideas of entrepreneurship education to reform the concept of innovation education in colleges and universities. The innovation entrepreneurship education should be put into the teaching targets, teaching contents and the course system reforming while emphasizing professional basic knowledge.

Concepts decides the acting ways, and acting ways determine the effect so the right concepts of education guide the right education action, which can effectively improve education effect. We should correctly understand the relationship between professional education and innovation entrepreneurship education, and innovate traditional, single professional education concepts with enhancing the students' social responsibility, innovation, entrepreneurial spirit and innovation ability as the core, reforming the personnel training mode and curriculum system as the key point to set up the educating ideas which combine professional education with creative education, and creative education will be penetrated into the whole process of professional education.

3.2   Teaching Methods and Examination Way Will Be Reformed

Educating methods will be further reformed and strengthening innovation and entrepreneurship education contents will be strengthened to cultivate students' critical and creative thinking, inspire them to carry out innovation and entrepreneurship. The creative elements will be introduced and practical teaching modules meeting the demand of social talent will be provided based on the leading edges of profession.

Students' innovative ability will be improved and students' autonomous learning abilities, problem analysing and solving abilities, team cooperation and social activities will be focused. Explore the new mode of "teaching and competing" which closely integrates the teaching link with students participating in various professional and technical skills contests. The teachers and students' concepts of teaching and learning will be changed and student's main body status will be highlighted. Teaching will be expanded to the workshops, stores, fields, open education. We continuously reform the teaching methods, and effectively use network teaching resources to encourage teachers and students to flexibly learn using online.

We will reform examination contents and modes, pay attention to the learning process, explore the non-standard answer test, pay attention to check the students' invention and innovation, internship and practice, practicing experience, the operating ability to get rid of disadvantages of "focusing on the theories and ignoring practice".

3.3 Strengthening the Curriculum System

According to the three key links of improving abilities, inspiring enthusiasm and creating conditions, we further define innovation entrepreneurship education goals and requirements to target at the needs of all students to carry out general education. The training education will be carried out for students with innovating intention and incubation and support conditions will be provided for entrepreneurial students.

"Entrepreneurship Basic" should be included in the talent training scheme and all students must accept general education of entrepreneurship. At the same time a entrepreneurial society will be set up in which regular discussions and activities will be organized, and the related training education will be carried out. The construction of incubation environment and support security efforts are to be strengthened. The entrepreneurship teaching and research section will be strengthened, and the whole process of entrepreneurship education will be systematically studied to compile the text books of the innovation and entrepreneurship training.

3.4 Optimizing Structures of Teaching Force

Entrepreneurial class teachers may be teachers with vocational education quality and entrepreneurship instructor qualification is required. Successful businessmen, entrepreneurs, venture capitalists and talented people in all walks of life can be hired to undertake to teach related courses.

Establish the staff with full-time and part-time teachers combined and the innovative undertaking part-time tutor database. Strengthen entrepreneurial teachers training system with innovative undertaking innovation consciousness and the ability of entrepreneurship education as the important contents of teacher training.

3.5 Strengthening Creative and Entrepreneurial Practice and Training

First of all, we enhance the enthusiasm of teachers' instructing students to conduct innovative entrepreneurial project based on majors with cultural projects of science, technology and culture and "project guide training" as the guide of practice teaching concepts. Second, contests of college students' innovative entrepreneurial class are held or supported to provide a platform for students to turn their creative ideas into reality products. Besides, the second classroom activities are carried out vigorously to support for students to set up innovation business associations, business clubs, entrepreneurial salons, hackers' space. We regularly organize innovation entrepreneurship lectures, forums, and conduct innovative business practices. Social, corporate training resources are integrated for college students with entrepreneurial intention and training needs to provide professional training for free. Finally, we make full use of various media to widely publicize the policies, measures and achievements of the authorities promoting innovative entrepreneurial projects.

Train the student to achieve the "three abilities":

First, having the ability of keen insight. In practice, students are trained to actively explore new business opportunities according to the actual demand and actively perceive for development projects of good prospects in the market.

Second, having a solid professional knowledge and the Entrepreneurial professional skills. Students should realize that the solid professional knowledge play a crucial role in the process of enterprise development.

Third, having the spirits of hard work and perseverance. The process of entrepreneurship is not overnight thing but encountering a lot of unexpected difficulties and obstacles, so encouragement and perseverance are needed to achieve success.

3.6 Innovation Entrepreneurship Education Pattern

A more specific innovation entrepreneurship can be built according the research above, as shown in Fig. 1.

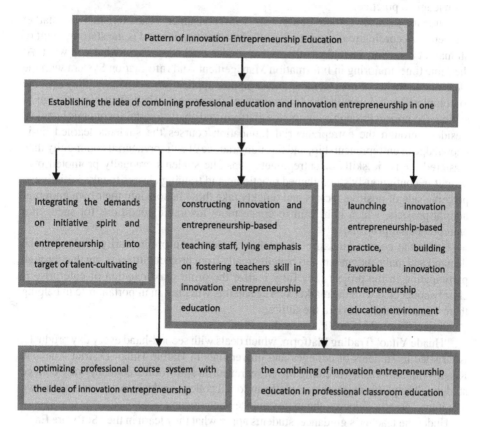

Fig. 1. Pattern of innovation entrepreneurship education

4   Practical Cases

Entrepreneurship education in colleges and universities should cultivate the students' innovative spirit and entrepreneurial thinking and entrepreneurial qualities under the background of innovation-driven development strategy, and students are enabled to develop in the fields of society, economy and culture with the professional knowledge to promote social and economic development.

Teachers in Harbin Huade University actively participate in "Heilongjiang Province College Students' Innovative Entrepreneurial Training Project" and constantly sum up experience to continuously explore new innovative start-ups.

Project One: The study of the market demand analysis based on creative products.

Instructors guide students to gradually master the methods of market research from the perspective of market demand. Students of the project are from the major of Information Management And Information Systems Professional, and they already have the corresponding professional knowledge and obtain precious opportunities to apply knowledge to practice.

Students themselves are consumers with consumer characteristics of market segments. According to the point of view of marketing, demand is the starting point of all market behavior, so students themselves understand more about what you want. At the same time, majoring in Information Management And Information System with the ideas and thoughts of management and computer application technology, they are able to complete such an analysis and research of the project. They even can manage to achieve the project through e-commerce platform and obtain a certain economic benefits. Besides, through the entrepreneurial foundation courses the students learned basic knowledge of entrepreneurship, and through practice of entrepreneurial community they mastered the basic skills of entrepreneurship. The students gradually promoted own insight in continuously strengthened practicing and training classes to take initiative to perceive side of positive factors and quickly take these factors into their own entrepreneurial projects, which is helpful to create more favorable conditions for successful entrepreneurial projects.

Under the correct guidance of teacher, students can preliminarily analyse social demand and make judgments and benefit greatly from project. Students say that through participating in the project they fully realize the profit maximization of enterprise is closely connected with the market survey and this will play an important role in helping them complete their work in the future.

Project Two: "Huade Yitao" Trading Platform.

"Huade Yitao" Trading Platform, which deals with second-hand every day products, is targeted to students of colleges and universities in Harbin Hulan District Xueyuan Road. With its website revealing the entrepreneurial spirit of the contemporary college students, The platform provides a platform for all college students to deal with their own idle items, greatly reducing the extravagance phenomenon.

Under the teacher's guidance, students apply what they learn in the "Software Engineering" in combination with software of website development to completing the transaction platform construction step by step. The students deeply feel that completing the project process in fact is the process of entrepreneurship, during which they realized the importance of teamwork, and the necessity of collaboration, and at the same time they learned the knowledge outside the books to enrich the experience of life.

What the students have achieved is closely related with the school's deepening reform of school education mode and strengthening practice and training innovative undertaking. Students of the project were selected into the project team because of the excellent performance in the practice link of "Software Engineering". They can use the knowledge learned in practice lesson, combined with teachers' innovative business ideas, to creatively complete each task, so that they get the exercise and achieve good results.

Project Three: Campus Environment Protection And Waste Recycling

Contemporary college students focus on how humans and the environment develop harmoniously, and it is a problem needed to be resolved when people can obtain the resources constantly from the environment. Campus environment resources is also a concern for the college students. It is the core of the research project plan how to effectively use campus resources without polluting the environment, and save cost.

After the project teacher's patient guidance, the students develop their own advantages and involve themselves in the project practice, so that they have created such a platform on which students timely exchange spare items for items deeded without leaving the campus. The platform improves the campus environment and brings income to the students. All the students think students' wide involvement and social influence are given priority to and the platform's radiating and driving function will be played. The project should be widely publicized and promoted to effectively explore public environmental benefits and social benefits. Through the project practice, every student has accumulated rich social experience and got full development.

The achievement of the project embodies the guiding effect of the ideas of innovation and entrepreneurship. Through guiding students to find valuable things around so as to deeply recognize the concept determines the way of action, action mode determines the effect in the new economic era.

Through innovative entrepreneurship training program of practice, the teachers and the students have accumulated a lot of experience, the students are in the application of professional knowledge and social experiences of growth has got great progress.

5 Conclusion

Strengthening innovation and entrepreneurship education is one of the important tasks of the future development of higher education reform. Innovation and entrepreneurship education is a challenge to traditional education concept, the innovation entrepreneurship education into the professional teaching, enables students to gain professional knowledge, and accept the innovation of entrepreneurial education, which can effectively help students to set up the correct view of entrepreneurship.

Professor Jeffry A. Timmons of Babson College of the US thinks that entrepreneurship education should focus on setting entrepreneurship genetic code for future generations to create the most revolutionary business generation as its basic value orientation of the prospective entrepreneurship education concept, it is essentially a business oriented revolution education innovation of human resource development.

Innovation entrepreneurship education is a long-term process and the impact is huge. Colleges and universities explore step by step innovative business mode combined with the actual situation of the school to further promote the university innovation entrepreneurship education. Continuously developing and improving entrepreneurship education service system make the practice of entrepreneurship education not become a mere formality to really promote the effective implementation of the student entrepreneurship practice activities, promote campus smooth transformation of scientific and technological achievements, and ensure the smooth of creative education.

Acknowledgments. This work is supported, in part, by key program of Heilongjiang Education and Science during the 12th Five-Year Plan period; the Grant numbers are GJB1215026 and GJB1215025.

References

1. Liu, C.: The analysis of carrying out the entrepreneurship education under the perspective of innovation-driven development strategy in colleges and universities. J. Liaoning Econ. Manage. Cadre Inst. J. Liaoning Econ. Vocat. Technol. Inst. **01**, 63–67 (2015)
2. Zhou, Z., Wang, D., Huang, Y.: How entrepreneurship education combine with actual working experience. Manager J. **07**, 260–261 (2013)
3. Lan, W.: The Research of College Computer Professional Students' Innovative Entrepreneurial Education Model. Southwest Jiaotong University, pp. 31–39 (2011)
4. Hao, H., Shuzhexing, Yu, J.: The research of entrepreneurship education resources platform construction engineering in colleges and universitie. China Mod. Educ. Equipment **09**, 141–143 (2015)
5. Ni, T., Gu, X., Cao, P.: Innovative talent training under 3 + 1 mode computer professional. Mod. Comput. **11**, 36–38 (2013)
6. Chao, H.: The Exploration and practice of cultivating the ability of students' innovative entrepreneurial computer science with Yu Lin Normal University as an example. Sci. Technol. Inf. **13**, 164–165, 180 (2013)
7. Zheng, Y.: The analysis of the ways of maker education in american colleges and universities. Open Educ. Res. **03**, 21–29 (2015)
8. Fang, Y., Wang, X., Xi, L., Chen, Z., Fang, L.: Practice teaching reform of computer major with innovation ability training as the core. Comput. Educ. **09**, 1–8 (2011)
9. Gao, H., Qiu, Z., Wu, D., Gao, L.: Research and reflection on teaching of C programming language design. In: Wang, H., Qi, H., Che, W., Qiu, Z., Kong, L., Han, Z., Lin, J., Lu, Z. (eds.) ICYCSEE 2015. CCIS, vol. 503, pp. 370–377. Springer, Heidelberg (2015)

Research of "Social Network Database System" Based on Flipped Classroom

Zhi-yong Luo[✉], Peng Wang, and Guang-lu Sun

Computer Science and Technology College,
Harbin University of Science and Technology, Harbin 150080, Heilongjiang, China
luozhiyongemail@sina.com

Abstract. Social network database system is a computer professional basic course. It is widely opened in engineering universities. However, teaching student to understand esoteric network database theory and master practical technology are a difficult problem. In this paper, we combine the prevalent at home and abroad flipped classroom teaching philosophy, and designed a model for teaching computer science students to learn social network database system. The model uses a new generation of information technology as a platform, and takings into account the flip classroom theory. This model fully embodies the "teacher assisted, student independent learning "teaching thinking, and effectively stimulates student interest in learning. The practical results show that the teaching model can enhance student achievement and the ability to solve practical problems.

Keywords: Flipped classroom · Reversed classroom · Teaching model · Social network database

1 Introduction

With the development of cloud computing, mobile Internet and new Internet technologies, information technology plays a more and more important role in the process of teaching and learning [1]. We use a new generation of information technology in education and teaching activities, not only can promote the reform of traditional teaching activities, but also can provide technical support for the teaching mode to the efficient [2]. According to the spirit of the document "Ten years development plan of education informatization (2011–2020)" [3], the computer science in every engineering college uses the latest information technology to establish teaching mode of "students autonomous learning and teacher assisted learning". Through the use of heuristics, discussion, inquiry and other teaching methods can development evaluation system and improve the quality of teaching. Flipped classroom teaching mode integrates information technology with traditional teaching to realize the requirements of development planning.

"Social network database system" is a practical application of the course, it has a strong theoretical and technical practice [4]. Therefore, in the course of teaching activities, it is necessary to explain the boring database theory, and strive to improve the ability of students to use these theories to solve practical problems. Applying the flipped

© Springer Science+Business Media Singapore 2016
W. Che et al. (Eds.): ICYCSEE 2016, Part II, CCIS 624, pp. 45–52, 2016.
DOI: 10.1007/978-981-10-2098-8_7

classroom teaching model to the traditional "social network database system" teaching process, not only can improve the efficiency of teaching, but also can improve the students' self-study ability.

2 Flipped Classroom

2.1 The Connotation of Flipped Classroom

Flipped classroom is that: before class, teachers use the network platform to upload teaching videos and other learning resources, students need to download and watch the teaching resources, through finding the relevant information to complete and submit to the understanding of knowledge points; in class, teachers will be teaching activities, such as heuristic, discussion and inquiry, to complete interactive communication and collaborative inquiry with students in order to achieve the purpose of strengthening the knowledge and skills training. Part of literatures [5] calls the teaching model "upside down classroom".

2.2 Research Status of Flipped Classroom

Flipped classroom was first proposed and applied by the United States "forest park" high school chemistry teacher Sams Aaron and Bergmann Jon [6]. Subsequently, they found that the teaching model is much better than the traditional in teaching effect. This kind of teaching mode is very quick in the United States and even the educational circles of the developed countries. In 2011, flipped classroom was rated as a major technological change in the classroom teaching by the Canada's "global mail" magazine.

At present, our country education scholars in this kind of teaching mode research are also more and more. On the basis of studying foreign teaching cases, Zhang Jinlei putted forward several flipped classroom teaching models; Zeng zhen discussed the features and common problems of the flipped classroom, and gave typical examples of the success of the individual teaching; Ma Xiulin turned the flipped classroom into the teaching of university information technology public course, and concluded that the teaching mode was helpful to master of knowledge points and improve the students' skill level. In a word, from the current literature, the research and application of the flipped classroom in our country mainly focus on the education of primary and secondary schools. In engineering colleges and universities, especially in the education and teaching activities for the specific courses of computer science, the flipped classroom results are less, and these require the majority of educators to study and supplement.

3   The Teaching Model of the Social Network Database System in the Flipped Classroom

3.1   Model Design

On the basis of previous studies [7, 8], the author combined "social network database system" course and the characteristics of the students of this specialty, and identified the reform of the teaching model as shown in Fig. 1.

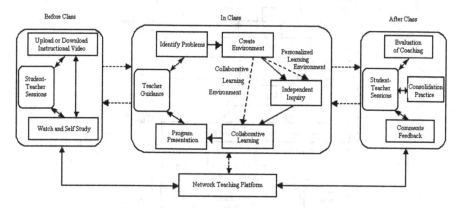

Fig. 1. The teaching model of social network database system

Figure 1 divides the teaching activities of the course into three parts: before class, in class and after class. Before class: teachers and students have their own different teaching tasks, teachers is mainly responsible for the induction of teaching materials, focus on teaching video recording, through the network teaching platform to upload to the network environment; students log on to the network teaching platform, download and self-taught learning materials, summarize the problems, timely communicate with teachers online, digest the relevant theoretical knowledge of network database technology. In class: teachers help students to complete the development of specific social network database system creating case, identify specific issues and create an independent inquiry, exchange of learning and the opportunity to show the program for each student. After class: teachers use the network teaching platform to complete the coaching of students' evaluation; students give timely feedback and teachers further urge students to review, consolidate the mastery of knowledge.

3.2   Network Platform Construction

The network teaching platform structure adopted by the "social network database system" course teaching reform is shown in Fig. 2. The structure is a B/S model, which includes three levels, namely the presentation layer, function layer and data layer. Among them: the function layer is composed of the teacher's function interface and the student's function interface, it is the core of the teaching platform, and it is also the technical guarantee for the successful implementation of the flipped classroom, it is used to complete the function of information transfer between teachers and students; the data layer provides services for the functional layer, it is to store the flipped classroom used in all kinds of teaching resources, including the basic information of students, students learning records, teachers and students online exchange records, teaching resources, video information and exercise answers, etc.; the presentation layer is the way of using teaching platform for teachers and students, mainly through the WEB browser to log on.

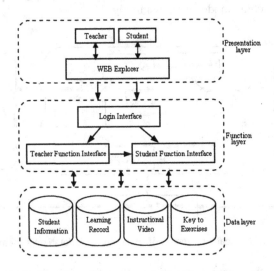

Fig. 2. Network teaching platform structure

3.3   Teaching Process Design

From Fig. 1 we can see that "social network database system" curriculum implementation "flipped classroom" teaching methods include three aspects, respectively is: design before class, design in class and design after class.

3.3.1   Design Before Class

"Design before class" is the prerequisite for the implementation of the curriculum teaching reform, so this stage requires both teachers and students to use their spare time to make the necessary pre-class preparation. Teachers should sum up the teaching materials needed in the next class, provide some video information, PPT courseware, knowledge point of the list and the necessary reference books and other resources, upload to the network environment through the network teaching platform, and give the necessary instructions. The teaching resources provided by the teachers are generally divided into two levels, namely, the basic level and the extended class level, in order to meet the different requirements of the quality of students. In the teaching materials, the video data is the core and request the teacher to make 1 to 3 video data according to each class the teaching goal, each video data introduce a web database development technology of knowledge or introduce a database development case, and configure a key point for documentation, In order to facilitate the students to study independently. Students according to the video information provided by the key points of the configuration documentation, combined with other supplementary information published by the teachers to learn. Students summarize the knowledge point of learning, record the problems found, and collect these records to the team leader. On the day before the class, the leader will submit these records to the teacher, accept the teachers' assessment and analysis, determine the course of teaching in the case and focus for the teacher.

3.3.2   Design in Class

"Design in class" is the key to the implementation of the curriculum teaching reform, students and teachers are required to closely cooperate, give full play to the "flipped Classroom" of the new teaching ideas. Therefore, the stage is divided into the following four steps, this article assumes that each course for 90 min.

(1) First, teachers use 15 min to summarize the emphasis and difficulty of the curriculum teaching, inform the student's learning situation, give the assessment score, display the score rules, arrange the teaching case of this course, and give the final goal of the case.

(2) Students complete the task of teaching the case independently and this takes about 50 min. During the period, the students can put forward some specific problems in the development of the database to the teachers, and the students will be guided by the teachers and complete independently.

(3) Students display their program works and it is expected to take 15 min. The student is the speaker and will fully display the program works that he developed, the student is the speaker and will fully display the program works that he developed, introduce the key commands and the realization process of the technology. Teachers and other students involved in the evaluation and talked about related technology; determine whether the work can be optimized and how to optimize. In order to further let the students understand the network database development technology and its own shortcomings, to be improved.

(4) Teachers sum up the development of this course and it is expected to take 10 min. According to the development of the course and the performance of the students, the teacher summed up the course in detail, Teachers to further enhance the teaching focus of this course, praise the excellent students, criticize the poor students, give the assessment results of this course and the time planning of the next course.

3.3.3   Design After Class

"Design after class" is a solidification process that students further understand and master the development of network database technology, so it is a very important link. The design stage is mainly divided into three aspects: feedback, assessment and coaching exercises.

(1) Feedback. In this part, students complete of the curriculum evaluation, gives the advantages and shortcomings as well as suggestions for improvement, will these opinions through the network teaching platform to upload to the teachers; after the summary

of the analysis, teachers complete the optimization of the curriculum to be carried out in the future.

(2) Assessment. In this part, students will finish the unfinished work that should be done in class, and upload these to the network environment; teachers will download these reviews and be uploaded to the evaluation of students, students further understand the relevant technology of network database development and knowledge in coaching teachers.

(3) Coaching exercises. In this part, the teacher will upload the full knowledge of this course exercises, students will download the information to learn, to further strengthen and consolidate the knowledge points.

4   Analysis on the Effect of the Teaching Model

4.1   The Experimental Process

The teaching reform of the course is to use the traditional class and the experimental class simultaneously to carry on the teaching method. The teaching object is the students of grade two in computer major, and the number of students in each class is 30. A questionnaire survey was conducted before the teaching, and the results were shown in Table 1.

Table 1.   Statistical table of students' computer basic quality

Prefect object	Students' number	Proportion men and women	Proportion of having computers	Proportion of using network	Proportion of programming experience	Used DB experience's proportion
Traditional class	30	22:8	90 %	93.3 %	66.7 %	10 %
Experimental class	30	21:9	86.7 %	96.7 %	73.3 %	13.3 %

Table 1 show that students in the traditional and experimental class have the same theory and skill level of the social network database system.

During the course of two years teaching, the author used traditional teaching methods to teach traditional class and used flipped classroom teaching methods to teach experimental class. Final examination results, questionnaire, teacher's own feelings and other aspects show that using flipped classroom teaching mode, the students are superior to the traditional teaching methods in the aspects of mastering the theoretical knowledge, skills, technology and learning interest.

4.2   Performance Comparative Analysis

Compare the academic performance of the traditional class and the experimental class from high to low, and get the results of the comparison chart as shown in Fig. 3. By calculation, the average score of the traditional class was 71.6, the average score of the experimental class was 75.23. The average score of the experimental class is 3.63 points higher than that of the traditional class.

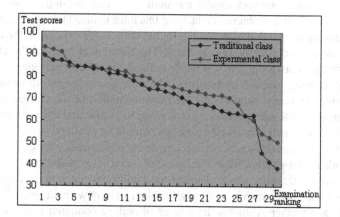

Fig. 3. Traditional and experimental classes' scores comparison chart (Color figure online)

Compare the academic performance of the traditional class and the experimental class with percentage of score interval, and get the results of the comparison chart as shown in Fig. 4. Figure 4 shows: The excellent rate of the experimental class is higher than that of the traditional class 10 %, the good rate is almost equal, the average rate of the experimental class is higher than the traditional class 13.34 %, and the passing rate of the experimental class is decreased by 20 %. In short, the total score of the experimental class is significantly better than that of the traditional class.

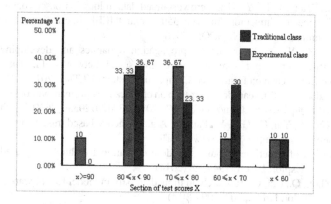

Fig. 4. The people numbers of each score section comparison chart (Color figure online)

Figures 3 and 4 show that, based on the flipped classroom teaching mode is more conducive to students to master the theoretical knowledge of the social network database system and improve their skills to solve specific problems.

5   Conclusions

If the course of "social network database system" is carried out in the mode of flipped classroom teaching, the problems of lacking teaching hours, lacking comprehensive cases, and lacking thorough understanding for the knowledge existing in the traditional teaching mode can be effectively overcome. The theoretical basis of the curriculum reform is "flipped classroom", which really realizes the educational idea of students' autonomous learning assisted by teachers. The new generation of information technology provides technical supports for the implementation of the curriculum reform. Teachers can upload teaching resources according to course and urge students to self-study. The learning method is also flexible, and cannot be confined to the classroom. It further strengthens the curing process in classroom communication for the development of the network database technology. The course of social network database system under Flipped classroom teaching mode changed the traditional concepts of education and learning of computer. It puts forward a new problem and a new challenge to the teachers and students and further promotes the change of college computer teaching toward a more effective way. It has provided a certain reference for the vast number of educators.

Acknowledgements.   Supported by Heilongjiang Provincial Education and Scientific Projects (No: GBD1211026), the Ministry of Education's Humanities and Social Science Project No. 11YJC740048, Scientific planning issues of education in Heilongjiang Province No. GBC1211062.

References

1. Li, J., Cai, Z., Yan, M., Li, Y.: Using crowdsourced data in location-based social networks to explore influence maximization. In: The 35th Annual IEEE International Conference on Computer Communications (INFOCOM 2016)
2. He, Z., Cai, Z., Wang, X.: Modeling propagation dynamics and developing optimized countermeasures for rumor spreading in online social networks. In: The 35th IEEE International Conference on Distributed Computing Systems (ICDCS 2015)
3. Ministry of Education. Ten years development plan of education informatization (2011–2020)[EB/Ol.], May 2012. http://www.edu.cn/zong_he_870/20120330/t20120330_760603_3.shtml
4. Wang, Y., Cai, Z., Yin, G., Gao, Y., Pan, Q.: A game theory-based trust measurement model for social networks. Comput. Soc. Netw. (2016)
5. Lin, Q.: Design of experiment course "modern educational technology" based on flipped classroom. Res. Explor. Lab. **1**, 194–198 (2014)
6. Zeng, M., Zhou, Q., Cai, G.: Research on flipped classroom model for software development courses. Res. Explor. Lab. **2**, 203–209 (2014)
7. Luo, Z., Qiao, P., Qin, Z.: Study on the teaching system reform of network database in engineering university. Sci. Technol. Manag. **1**, 127–129, 135 (2011)
8. Luo, Z., Yining, X., Wang, L.: The research and reflection on the teaching reform for the system of network database. J. Jiangsu Inst. Educ. (Nat. Sci.) **3**, 100–103 (2009)

Research on Interactive Simulation Experiment Platform and Remote Simulation System Under Web Environment

Bing Zhao[✉], Zhifang Wang, Jiaqi Zhen, and Erfu Wang

Department of Lectronic Engineering, Heilongjiang University,
Harbin, Heilongjiang, China
zb0624@163.com

Abstract. This paper brings forward a method of experimental teaching reform of universities and colleges, which establishes an interactive simulation experiment platform with MATLAB simulation software. The platform uses visual man-machine interface, which avoids a lot of programming and debugging work, and improves students' learning enthusiasm. Simultaneously, it implements remote simulation based on JAVA under web environment, which is convenient for network teaching, and improves flexibility and practicability of the simulation experiment platform.

Keywords: Experiment platform · Interactive simulation · Web · Remote simulation system

1 Introduction

Along with maturity of virtual technology, the interactive simulation experiment and remote simulation system present an attractive application prospect. A lot of education institutions at home and abroad have carried out research and construction of the interactive experimental platform, which has become a hot point in research of education teaching, especially applying in experiment teaching of electronic information engineering, communication engineering and other information-type disciplines [1,2]. Domestic universities and colleges have carried out research on primary-secondary arrangement and emendation of experiment content, and division and engagement of experimental and theoretical teaching, and applied them into the specific experiment teaching. The experimental teaching of information-type courses is various in content, covers broad scope, and requires a lot of labor power and money, therefore, establishment of interactive simulation platform and remote simulation system under web environment based on MATLAB, Java and other software becomes one of the important methods for the reform of experimental teaching in colleges and universities [3–5].

W. Che et al. (Eds.): ICYCSEE 2016, Part II, CCIS 624, pp. 53–58, 2016.
DOI: 10.1007/978-981-10-2098-8_8

2   Interactive Simulation Experiment Platform

MATLAB simulation software is one of the most commonly used numerical calculation software. It has powerful calculation function, convenient drawing function, and various tool bags. In order to carry through visualized information communication between users and computers, MATLAB provides Graphical User Interface (GUI), and the simulation experiment platform based on GUI can complete experiments under different courses with the programs written in computer in advance. It doesnt require too much expensive experiment equipment and device, which saves experimental materials and personnel, solves insufficient experiment funds, and is convenient for update and emendation of experiment content. Secondly, MATLA GUI is a method with human-computer interaction. Its biggest advantage is that students only face with intuitive graphical interface in process of experiment, don't need to mechanically memorize a lot of programming statements, but carry out the commands of choose, calling file, starting program, or other operations through the enter with keyboard and mouse. Figure 1 is static picture of Discrete Cosine Transform (DCT) experiment designed with GUI.

After trial, students' feedback is good. They think experimental results can be clearly and intuitively seen from the experimental platform, and parameter modification is also very easy. It is easy to compare experimental effect under different parameters, helps them reduce the pressure from the emotion of not familiar with programming, and improves learning enthusiasm and teaching results.

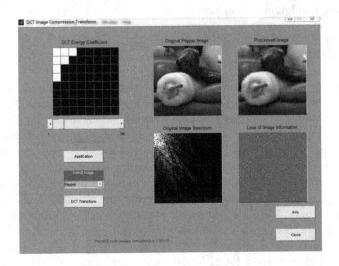

Fig. 1. Discrete cosine transform

3   The Remote Simulation System Under Web Environment

MATLAB not only has powerful calculation function, but also can be used to build remote simulation system under web environment. The toolbox in early version of MATLAB provides MATLAB Web Server tool bag to build web remote laboratory, while the new version of MATLAB provides Java Builder tool bag to complete remote operation [6]. The difference between these two sides is: the working mode of MATLAB Web Server works let students request data from Web browser to Web Server through Common Gateway Interface (CGI). CGI is an old server-side technology, and its biggest disadvantage is its efficiency is not high. When Web server receives a request related to CGI procedure, it needs to generate a new copy of procedure to operate. When there are a lot of requests, huge work load will occupy a large part of system resource, overwhelm server, and reduce its efficiency. Java Builder toolkit is able to avoid repetitive loading and initializing in data of MATLAB Web Server, and forms a Java assembly by packaging MATLAB function into one or multiple Java types with high efficient server technology Java Servlet. The method of packaging each MATLAB function into Java type can be invoked in Java application program. When Web server receives request from user to Servlet, server dispose request only with a new thread but not a new progress. Thread needs less resource than progress. Therefore, this will bring faster speed, as well as small load and higher efficiency of operation system.

Computers used by students are terminals, which do not need MATLAB software, and can easily control simulation experiment. All complex operation will be done in server provided by school or teachers. Teachers can change experiment content and add creative experiments according to course requirements.

4   Advantages and Targets of System

This experiment system aims at establishing comprehensive experiment platform with interaction function with simulation software, realizing long-distance simulation under web environment in this platform, and expanding experiment contents without increasing hardware equipment. This will not only improve students learning interests, cultivate their research ability, comprehensive application ability and creativity ability, and provide methods and basis on work in hardware design. Its advantages are:

(1) Abstract and difficult formula or conception can be changed to visible image with friendly interactive interface from text with simulation software. This will greatly improve class teaching quality.

(2) Avoid former fussy program input, output and limitation of debugging of previous simulation experiment, decreases fault rate of hardware device, reduce time of single experiment, and improves efficiency of experiment teaching.

(3) Long-distance simulation system eliminates limit of time and space of previous experiment. Students can do simulation experiment through network

at any time and in any place, and do not need to install simulation experiment in terminal.

(4) Before hardware design, the simulation service provided by this simulation platform offers function verification and effect analysis, which improves design efficiency and reduces device consumption.

After being completed, the system will be applied in experiment teaching of information series curriculums, the following targets is predicted to be got:

(1) Reform traditional experiment teaching method. With the carrier of simulation platform, this system simulates real simulation environment and traditional experiment device, which helps to integrate experiment curriculum contents of information series, and get original process of knowledge.

(2) Establish a man-machine interactive platform with friendly interface, powerful function and easy operation, and increase students learning interests. Conveniently update and add experiment contents with experiment platform, which saves experiment space and saves cost.

(3) Increase the establishing rate of comprehensive, designing and creative experiment, enlarge opening scale and layer of experiment, and improve students innovation level and scientific research awareness.

(4) Assist class teaching, open network e-classroom in the internet, provide long-distance simulation experiment under web environment, and enhance flexibility and practicability of simulation experiment platform.

5   Class Practice

Using existing equipment in laboratory and MATLAB to establish a remote simulation experiment platform, I conducted an open experiment among 120 students who choose the spread spectrum communication course in communication engineering major at the end of last term. The experiment has 12 lessons, divided into 4 parts, students can choose to use the simulation platform or to laboratory. There are 98 students use the remote simulation platform. I organize a questionnaire survey among the 98 students, questions of the questionnaire are: Do you think remote simulation experiment platform can help students better accomplish experiment content and help to arrange experiment time and reduce the course burden, and play a favorable promoting function in your study? The results are shown as the Fig. 2.

Results show that more than 81.6 % students hold supportive attitudes, students approval rating has reached 94 % on question 2. Most students believe that the remote simulation platform can make full use spare time and improve the learning efficiency. At the same time, from the feedback of the students we also learned that senior students learn specialized courses mainly, students choose courses are not same, spare time are different and scattered. Therefore, it is difficult to arrange a uniform time for experiments, the experiment process is too long which is not conducive to content convergence. So remote simulation platform can facilitate students greatly and stimulate students learning enthusiasm.

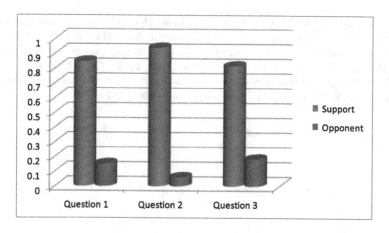

Fig. 2. The results of questionnaire survey (Color figure online)

6   Conclusion

Establishing interactive simulation experiment platform and long-distance simulation system with simulation software like MATLAB largely increases flexibility of experiment teaching, overcomes time and space limit of experiment curriculum, facilitates experiment terminal operation, decreases requirement on students computer device and system environment. The platform and system is suitable to network remoteness teaching very much, which provide long-distance interactive simulation experiment under web environment, send data to server with Internet, and then return data and figure results to terminal with HTML format with image display function of simulation software. This system does not only stimulate students learning enthusiasm on experiment curriculum, improve their learning efficiency, but also lays a good foundation for their work in software and hardware design in the future.

Acknowledgements. This work is supported by 2015 Heilongjiang University New Century Education and Teaching Reform Project (NO.2015B76). Many thanks to the anonymous reviewers, whose insightful comments made this a better paper.

References

1. Wang, J., Yuan, Z.: The application of virtual simulation technology in the teaching of single-chip microcomputer course. Electron. Des. Eng. **24**(1), 45–47 (2016)
2. Wen, H., Chang, S., Cai, Z.: Research and practice of FPGA experiment virtual simulation platform construction. Electron. Technol. Softw. Eng. (2), 86–87 (2016)
3. Dai, Y., Qiu, D., Zhang, B., Xiao, W.: Design of simulation experimental platform of phase-shift full-bridge converter based on Matlab. Exp. Technol. Manag. **28**(5), 86–89 (2011)

4. Jin, W., Gao, R., Wang, Y.: New strategy to develop web based remote simulation system using Matlab Java Builder. Microcomput. Inf. **27**(4), 204–206 (2011)
5. Yao, M., Zhao, M., Xing, L.: Neural network simulation experiment design based on Matlab and VC++. J. Electr. Electron. Educ. **29**(3), 77–80 (2007)
6. Cai, Y.: Design of distance education system for mathematics based on Java Web and Matlab Builder JA. Exp. Technol. Manag. **29**(1), 83–85 (2012)

Study of Flipped Classroom Teaching Mode Suitable for China's National Conditions

Fang Yin[1(✉)] and Rui Wu[2]

[1] School of Computer Science and Technology,
Harbin University of Science and Technology, Harbin 150080, China
13936421412@163.com
[2] School of Computer Science and Technology,
Harbin Institute of Technology, Harbin 150001, China

Abstract. As a new teaching mode, Flipped Classroom has become the hot issue of teaching research and it has been accepted and praised highly by more and more researchers and schools. It changes teachers' dominant position in class and the relationship of teaching and learning between teacher and students, and students change passive learning into active learning. This paper deeply analyzes the essence and connotation of the flipped classroom, the development and practice of it at home and abroad, also existing problems in Flipped Classroom activities in our country were carried out to make the flipped Classroom to play an important role. Teaching practice verified that the Flipped Classroom teaching mode can maximum limit to mobilize students' study enthusiasm, improve the learning efficiency and it is an effective way of teaching.

Keywords: Teaching mode · Flipped classroom · Autonomous learning · Study enthusiasm

Flipped Classroom is that adjusting the time inside and outside the classroom to shift the learning decision from teachers to students, which may be understood as a means to increase interaction and personalized contact time between teachers and students and personalized contact time. With the development of information technology, the traditional teaching mode can't meet the requirements of students, so the Flipped Classroom mode originated from North America comes into our country and has become the hot topic for the education experts.

1 The Connotation and Essence of Flipped Classroom

As a new teaching method, flipped classroom is totally different from the traditional teaching mode that we always use. It is a completely reverse teaching idea. Comparing flipped classroom with traditional classroom that can help us deeply understand the connotation and essence of flipped class [1].

In traditional teaching, the teacher who controls the teaching process and the realization of teaching goals completely is the main part of the teaching activities. However, the students just passively accept new things and new knowledge from the teachers in

© Springer Science+Business Media Singapore 2016
W. Che et al. (Eds.): ICYCSEE 2016, Part II, CCIS 624, pp. 59–64, 2016.
DOI: 10.1007/978-981-10-2098-8_9

class. So the course teachers' knowledge level, the clarity of expression, the accumulation of experience and the teaching effect will directly decide the students' learning effect. Because students basically don't understand the knowledge the teachers speak before the class, students just dumbstruckly listen to a flow of words the teachers speak and take notes mechanically in the whole teaching process. We radically don't know how much the student understood and accepted. This is a passive learning process for students that students use a lot of time to read books, do their homework to deepen the understanding of the teacher speaking content after students are taught a lot of new knowledge in class.

In flipped Classroom the main body of teaching activity changes from teacher to students. According to the arranged contents, students use slides, video and other network resources to preview independently before class and evaluate themselves on the learning platform. In the learning process, students completely rely on personal ability to master the basis of new knowledge and the questions they met may be resolved by teacher and other students in class. So there is plenty of time for answering questions, group discussion and group practice interactive activities in class. As can be seen, according to individual ability and preference, students learn new knowledge in different ways before class and achieve the same level to pass the test on platform. In flipped class, teacher is changed from explainer to guidance and carries out a variety of flexible team activities to deepen understanding what are taught in class. For students, learning becomes active exploring process which not only arouses the enthusiasm of the students, but also improves the learning effect.

Through the above comparison it is shown that the essence of the flipped classroom is that students master the learning initiative and it can increase the interaction between teacher and students in the limited class time. Various kinds of interaction activities are guided by teachers to improve students' learning effect of new knowledge [2]. However, the way to achieve essence is the Flipping. The flipped classroom is not only flipping learning time, but also a variety of factors are flipped at the same time. Teacher's role is changed to the collective learning director, activities organizers. And the students' learning style is changed from passive learning to active learning. Through independent learning, exploring learning and collaborative learning before class, they build knowledge structure. The Flipped Classroom contains a large number of the communication between teacher and students and among students such as communicate on the Internet before class, group activities, the teacher answering questions for students and students exploring problems exists in the flipped class, which no longer just limited to answering questions after class, but also answering the preview question before class, discussion in class, even a debate. So the interaction between teachers and students has changed. The teachers' lesson preparation is changed. Because the students as the main body of teaching activity, teacher's teaching content and way need adjustment according to students' mastery degree of knowledge and problems in class. The above change inevitably requires teacher to have very solid professional foundation and professional work attitude.

2   The Development and Research of the Flipped Classroom at Home and Abroad

2.1   Foreign Research Status

The teaching model of flipped classroom originated from classmate teaching that founded by Harvard University physics professor Eric Mazur in 1991. The goal is to construct self-learning environment for students. This way is to broadcast knowledge to students through peer system by the teacher before class and they learn autonomously to master the basic concepts. Through the system to achieve the target of students really grasp the basic concepts, student know what is the key and difficult content in class. And through the teacher-students interaction, student complete the knowledge internalization. Experiments show that this method improves teaching effect significantly compared with traditional one. It has been widely adopted in many regions in the world. In 2000, The concept of flipped classroom was first proposed in the United States. That Salman Khan recorded video for cousin to learn math plays a decisive role in the promotion and development of the flipped classroom.

2.2   Domestic Research Status

In our country, educators have also carried out a large number of related researches and practices on Flipped Classroom, and constantly create new models for our country's educational characteristics.

Zhang Jinlei [1] and other scholars build a more perfect model of the Flipped Classroom teaching model based on the study of American scholars according to the information technology and collaborative learning. It has played a guiding role in the research of our country. On the basis study results, Sun Jie further described the role of teacher and students in the Flipped Classroom and further improved the model. Even the relevant activities were carried out in primary and secondary schools such as Jukui middle school in Chongqing, Nanshan middle school in Shenzhen, the fifth middle school in Guangzhou, the first secondary school in Changle of Shandong province and so on. [2] They have introduced the Flipped Classroom model into teaching activities successively. And it is welcomed by teachers and students. The research and practice represented by Educational Technical College, Beijing Normal University, have been carried out in a large number of colleges and universities, and have made gratified achievements.

3   Problems of the Flipped Classroom

Although the Flipped Classroom has gained a keen concern and a lot of theoretical research and teaching practice have made great achievements, but there are still some shortcomings.

3.1 Following the Trend Too Much and Lacking Connotation

Now, due to the exploration of the Flipped Classroom model by experts and scholars caused a great uproar, which results that most colleges and universities carry out related work. But some practitioners do not really understand the essence and lead students to get less knowledge or skills through the reform of teaching mode and complained.

3.2 Making the Flipped Classroom Only Micro Video Production

The essence of Flipped Classroom is to increase the interaction between teachers and students and to increase the time of the personalized contact. Micro video is one of the important steps in this teaching method. It is the main way to learn before class for the students, and the quality of video directly affects the level of students' mastery of knowledge. But if it is regarded as the only way to flip, it is simply equaled to MOOC and micro lesson.

3.3 Personal Contact Between Teacher and Students is not Enough

The purpose of Flipped Classroom is to fully use class time for interactive activities, so teacher can provide each student individualized instruction according to their characteristics, fully satisfy. Now, due to the exploration of the model by experts and scholars caused a great uproar, which results that most colleges and universities carry out related work. But some practitioners do not really understand the essence and lead students to get less knowledge or skills through the reform of teaching mode and complained. But at present the domestic practices are mostly preview and discussion in class without perfect feedback system to track feedback, which can grasp the state of students learning and give them necessary guidance.

3.4 Network Platform is not Perfect Enough

As the support environment of the flipped classroom, its development level directly determines the effectiveness of the Flipped Classroom. Although since 2011, our country has begun to carry out the construction of education information, but lack of the effective policy system environment and mechanism which promote the development to the Flipped Classroom. At the same time, the lack of high-quality teaching resources, the poor ability of resource sharing, the poor infrastructure and so on still exist.

4 Establishing the Flipped Classroom Teaching Mode

Based on the problems it is necessary to deepen the understanding of its essential connotation, and to study the appropriate practice methods of the flipped classroom teaching mode according to the characteristics of China's national conditions.

4.1 Analyzing the Characteristics of Our Country's Education, Explore the Characteristics of the Flipped Classroom Teaching Mode

The flipped classroom is originated from the abroad, and it is still in the research stage and the preliminary practice. There must be a lot of content which is not suitable for our country. We should inherit and accept the content with criticism. In order to let the flipped classroom can really play a positive role in the process of teaching, education experts, researchers should deeply explore the connotation and the essence of the flipped classroom, according to the current situation of our country's traditional classroom teaching, and research a really suitable for the flipped classroom teaching mode with ourselves' teaching characteristics, to reach that the teaching activities under this model can really mobilize students' learning initiative and enthusiasm so that all levels of various characteristics students are able to get effective guidance, to exploit their maximum potential.

4.2 Improving Teachers' Own Quality and Professional Ability, Laying the Foundation for the Flipped Classroom

Focusing on the flipped classroom teaching mode about enhancing teacher-students interaction, and promoting into the respect for the personality development of students, all aspects of teachers' professional ability, practice ability, teaching ability of information technology was required of raising in the natural increase. Students' self-learning mainly relies on the information network platform, teachers to help students complete the self-learning link need to made a large number of course-ware and teaching video resources, at the same time, and to provide students with online self-test, and with students in the network platform to play a discussion, which will require that teachers must be with excellent ability and level. Under the flipped classroom, the teaching content is no longer old design, teachers need to adjust content and methods according to the students' mastery and a variety of ways of feedback from students, and to change the traditional "what I want to speak" for "students need what I say", which requires that teachers should be with good and rich knowledge and professional ability, strain at any time, and adapt to the needs of the students. The mode of teachers is much larger than the workload of the traditional classroom teaching, teachers must be with the spirit of dedication, and the devotion to teaching activities, truly to be the leading role of the beacon for students who explore knowledge.

4.3 To Strengthen the Construction and Improvement of the Network Platform

Most of the poor developed areas in China, especially in remote mountainous areas, can not provide enough hardware support for the construction of the network platform. In order to speed up the practice and application of flipped classroom, it is needed to speed up the popularization of information teaching, so that more students can benefit from the flipped classroom.

4.4   Truly Realizing the Students "Self-learning", "Individual-Culturing"

The flipped classroom emphasis on promoting students' self-learning, respecting for the individual development of students, in the teaching process implementation this theme must prosecute to the end. Each student on the understanding level of knowledge, acceptance rate, and the perspective of considering the problem being not identical, So in the teaching work teachers should consider the various types of student learning habit, according to the characteristics of each student take personalized guidance. Because of differences between students, in order to reflect the reasonable assessment about students, teachers must design a reasonable assessment mode of studying the knowledge of the present stage condition for students. The unified examination traditional form is not suitable for the teaching purpose of individual training. It is a necessary condition for the implementation of the concept of flipped classroom to look for an effective and comprehensive assessment method.

5   Conclusion

Flipped Classroom as a new teaching mode changes teachers' dominant position in class and the relationship of teaching and learning between teacher and students, and students change passive learning into active learning. It leads us into a new teaching mode and teaching practice verified that it can maximum mobilize students' study enthusiasm, improve the learning efficiency and is an effective way of teaching.

References

1. Jinlei, Z., Ying, W., Baohui, Z.: Introduction a new teaching model: flipped classroom. J. Distance Educ. **4**, 46–51 (2012)
2. Goodwin, B., Miller, K.: Evidence on flipped classrooms is still coming in. Educ. Leadersh. **70**, 78–80 (2013)
3. Chaoyng, H., Yufang, Q., Qi, C.: Inspiration of USA universities inverted classroom teaching model. Res. High. Educ. Eng. **2**, 148–151, 161 (2014)

Study on the College Politics Education Strategies and Methods in the Internet Plus Mode

Jiawei Ren[1], Lina Shan[1(✉)], and Xiaohui Meng[2]

[1] School of Reserve Officers, Harbin University of Science and Technology,
Harbin 150080, Heilongjiang, China
329382157@qq.com
[2] School of Computer Science and Technology, Harbin Institute of Technology,
Harbin 150001, Heilongjiang, China

Abstract. The flourishing development of Internet brings profound and lasting influences on the young students' growth. We use the complex network to analyze the characteristics of the information dissemination on Internet. We take the new requires for the educators, which occurred because of the students' pluralities, possibilities, playfulness and participation under the influence of Internet technology. We use the Markov chain theory to describe the transfer relationship among the four states including trust, agreement, participation, development of the college politics education patterns in the Internet Plus mode. In addition, we use hierarchical structure to calculate and offer suggestions for the distribution of the educators' work. We suggest five strategies including the conversational cooperation strategy, the service driven strategy, the O2O marketing strategy, the game participation strategy and the digital operation strategy. Then we perform the AHP method to find the best strategy.

Keywords: Internet Plus · Complex network · Markov chain · College politics education strategy · AHP

1 Introduction

When we refer to the fast historical evolution of the Internet Plus concept [1], which eventually becomes a government plan since it is put forward theoretically, we could see that derived from a theory of application and service, it has gradually become one of the most important motivating forces. It encourages the industrial innovation, promotes the cross-industry combination, benefits the social livelihood and accelerates the innovative development of the economy and the society in our country.

We could not deny that college politics education is an important part of the high school personnel cultivation. We consider that the colleges and universities should actively broaden their horizons to the upstream schools and downstream schools [2]. It is important for colleges and universities to blend in the historical trend of the economical and social development. In addition, we consider it necessary for the colleges and universities to perform analysis from the point of view of the country strategies.

© Springer Science+Business Media Singapore 2016
W. Che et al. (Eds.): ICYCSEE 2016, Part II, CCIS 624, pp. 65–77, 2016.
DOI: 10.1007/978-981-10-2098-8_10

Therefore, under the background of the fast development of the concept Internet Plus, the research topics about the it are with both theoretical and practical significance, such as what should we do to discover the characteristic changing in the young students and the new ways for spreading the college politics education, what should we do to conclude the basic work patterns of the college politics education in the new period that brought by the changing, what should we do to formulate the basic strategies and the specific methods of the college politics education to adapt to the Internet Plus mode.

2   Analysis on the Main Characteristics of the Young College Students Deeply Influenced by the Internet Development

In the book *Creativity and Education Futures: Learning in A Digital Age*, Anna Craft discussed the changing in the growth characteristics of the young students under the influence of the Internet technology. He concluded that the young students under the vital influence of the Internet are with four characteristics including pluralities, entertainment, possibility and participatory [3]. The four characteristics offer effective reference for deeply understanding the young students who are the main part of the college politics education.

Pluralities. The fundamental property of Internet is that it changes the inherent ways for communicating information in the human society. It broadens the selection pluralities for people when they need to make decisions. It costs little time and money to obtain the options of pluralities. Thus it will bring great pluralities in the young students' study and livelihood. Internet opens a door for the young students. The door not only leads them to more chances of exploring the world, but makes it easier for them to reveal multiple characteristic profiles. We should make adjustments in the college politics education to the changing. We should grasp the characteristics of the young students by means including broadening the working platform, enriching the working carrier, designing the working contents. Seizing the opportunities provided by the reform caused by the Internet Plus mode, we need to gradually update and complete the ideological and political work system. We should provide the young students with individualized resources and patterns that are effective, accurate and intimate by providing them with diverse services.

Possibilities. A variety of virtual games and original contents on the Internet and a variety of possibility thoughts caused by them transfer the sentence 'What it is' to 'What it could be'. The Internet technology and the digital media technology makes it easy to extract resources and complete missions, both of which are hard to realize in the conventional ways. In the conventional age, people respect and admire authorities. However, the respect and the admiration gradually disappear as the postmodern culture are spreading, the Internet age 2.0 is occurring and the authorities are breaking down. Every one can be a master node and an active self-media subject on Internet. Therefore, the conventional knowledge system, which are stable, changeless and requires continual learning, is no longer stable in the Internet age. The node groups on the Internet are updating and creating emerging knowledge everyday. The youth who grow up in the Internet age are with more skeptical spirit, critical spirit and independent spirit. It becomes a common phenomenon in the college politics education.

Playfulness. The post-95s young students who grow up with Chinese Internet has now become the main force in the colleges. They find plenty of entertaining experiences of their virtual self construction and accumulating experience in the virtual world in the online games, the social networks and the original contents. They remodel ego in the virtual world. The ego could to some certain extent release their id in the real world and make them experience the virtual existence of their superego in the real world [4]. The youth that grow up with virtual worlds on Internet show strong psychological dependence for games and entertainment. Entertainment has become some basic lifestyle and thinking habits of the young students. To some certain extent, their worldview and values are influenced by and depending on the entertainment.

Participation. There are various digital devices including cellphone, game consoles, laptops and pads provide the young students with Internet contacts and opportunities to high participation. The digital media technology and the Internet technology not only provide the young students with chances to act and practice by their entertainment, but also make it easier for them to be heard and discovered by others. We can know from the participation that the process of the college politics education is no longer the education and management work between the educators, the subjects, and the students, the objects. It reveals inter subjectivity. We should design a working system that is more scientific and use it to stimulate and develop students' active consciousness. In the process that the students fully participate, the educators should supplement active guidance and mental edification on the students in a Imperceptible way to achieve the fundamental purpose of the college politics education.

3   The Propaganda Ways of the College Politics Education in the Internet Plus Mode

Internet is a kind of classical complex network. Barabasi and Albert brought forward the scale-free network in the late 20th century [5]. It perfectly describes the property of Internet. We will discuss the features of the information spreading on the Internet based on the theory of the scale-free network to illustrate the importance of Internet. Furthermore, we bring some corresponding suggestions for the young students who are the main parts in the college politics education.

We use the software Pajek to draw a scale-free network with 100 nodes based on the scale-free network generating algorithm [6]. We use the nodes to represent the students and educators and the edges to indicate the links between them. We analyze the spreading ways for the ideological education and put forward some suggestions.

Firstly, we draw the links between students and educators in the situation that Internet is not popularized yet. We show the result in Fig. 1.

We can see from the figure that there exists few spreading ways for the college politics education. It is hard to develop the college politics education in the network that focused on the educators. We can see that when some ideologies spread from

Fig. 1. The interpersonal relationships before internet is popularized

educators to students, the topological distances are up to 5. We could observe the situation obviously when the nodes are numerous.

However, the appearance of Internet improved the situation. Based on this feature, we draw a network in Fig. 2.

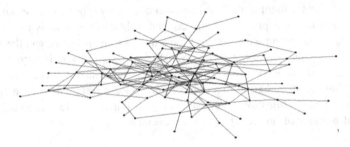

Fig. 2. The interpersonal relationships after internet is popularized

We can see from the figure that the topological distances reduces a lot. We consider that Internet brings new spreading ways for the college politics education. We find it regular that when a new node joins the network, it prefers to contact with the existing nodes with larger degrees. In addition, the relationships between the educators and students also derive from the relationships between subjects and objects to the relationships between subjects. Therefore, the links between students contribute to conducting the ideological education work. The newly built links increased the clustering coefficient of the network and correspondingly deepened the links between educators and students.

In the Internet Plus mode, we need to improve the college politics education into the following aspects. The educators should actively build more links with students. We can see from the figures that if we create new edges between the nodes with large degrees and other nodes, the degrees of these nodes will get larger. Then we can observe the Matthew Effect. The educators should try to find the groups that have few links with themselves and perform politics education on them. We can see from the figures that the nodes with large degrees should actively build links with the small groups that separated from the network. By this mean, we can easily build the relationship between a group and a educator. We should carry out more activities among students to promote the intercommunication between them. We create edges between the nodes randomly to reduce the average distance of the network and increase the spreading velocity of the college politics education.

4 The Basic Patterns of the College Politics Education in the Internet Plus Mode

We conclude from the deepening and evolution of the relationship between two subjects and the development and changing of the cognitive psychology of the educators and students that the process of the college politics education can be described as four basic interaction parts including trust, agreement, participation and development. We use the hierarchical-structure model based on the Markov-chain model to deeply analyze the college politics education patterns. We use the hierarchical-structure model to describe the relationships between the four basic interaction parts.

4.1 Determine the Basic Values

- The hierarchy $i = 1, 2, 3, 4$ respectively represents trust, agreement, participation and development.

  Quantity Distribution $(n_1(t), n_2(t), \dots n_k(t))$, where $n_i(t)$ denotes in the tth year the number of people belonging to the ith hierarchy in $n(t)$. It satisfies the equation

  $$N(t) = \sum_{i=1}^{k} n_i(t) \tag{1}$$

  Ratio Distribution a(t) = (a_1(t), a_2(t), ... a_k(t)), where

  $$a_i(t) = \frac{n_i(t)}{N(t)} \tag{2}$$

  Transport Matrix

  $$Q = \{p_{ij}\}_{k \times k} \tag{3}$$

  where p_{ij} denotes the proportion of people transport from i to j.
- Exit Proportion

  $$w = (w_1, w_2, \cdots, w_k) \tag{4}$$

  which represent the proportion of people exit every year.

- Exited Population

$$W(t) = \sum_{i=1}^{k} w_i n_i(t) = n(t) w^T \tag{5}$$

where t denotes the time of year
- Transfer Proportion

$$r = (r_1, r_2, \cdots r_k), r_i \tag{6}$$

which indicates the proportion of people transfer in i.

4.2 Establish the Basic Equations

We have the hierarchy structure basic equation [7]

$$n(t+1) = n(t)(Q + w^T r) + M(t)r \tag{7}$$

When the inward population and outward population of the system are approximately equal, we simplify the equation and have

$$a(t+1) = a(t)P = a(t)(Q + w^T r) \tag{8}$$

4.3 Establish the Stability Domain

We need a stable system structure so that we can use the transferred proportion to perform dynamic adjustment and achieve our final goal.

Based on the research and the data of the basic patterns of the college politics education in the colleges and universities in the previous years, we normalize the transfer relationship in every hierarchy. Then we have the transfer proportion matrix (Fig. 3).

$$Q = \begin{bmatrix} 0.8 & 0.4 & 0 & 0 \\ 0 & 0.6 & 0.4 & 0 \\ 0 & 0 & 0.4 & 0.3 \\ 0 & 0 & 0 & 0.8 \end{bmatrix}$$

If there exists r that makes a equal to $a(Q + w^T r)$, we call $a = (a_1, \cdots a_k)$ a stable structure. We have

$$r = \frac{a - aQ}{aw^T} \tag{9}$$

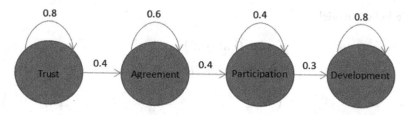

Fig. 3. The hierarchical structure and the Markov chain transition probabilities

When the equation satisfies the condition that $a \geq aQ \Rightarrow r \geq 0$, we can have the stability domain of the hierarchy structure.

We assume that there exists at least one i that makes $\sum\limits_{i=1}^{k} k_i$ less than 1. Then **M** is equal to inv(**I-Q**), we denote the row vector as m_i. Then

$$a = \frac{\sum\limits_{i=1}^{k} r_i m_i}{\sum\limits_{i=1}^{k} r_i \mu_i}, \qquad (10)$$

$$b_i = \frac{r_i \mu_i}{\sum\limits_{j=1}^{k} r_j \mu_j}, \qquad (11)$$

$$s_i = \frac{m_i}{\mu_i} \qquad (12)$$

We consider it the stability domain of the hierarchy structure system if it belongs to the following ranges.

$s_1 = (0, 0, 0, 1), s_2 = (0, 0, 0.4, 0.6),$

$s_3 = (0, 0.286, 0.286, 0.428), s_4 = (0.240, 0.240, 0.160, 0.360)$

S_4 is one of the solutions of the stability domain, we can consider it as the population proportion of the hierarchy structure.

4.4 Dynamic Adjustment Based on the Transferred Proportion

Our final goal is to reach or get close to the given deal hierarchy structure a^* as soon as possible by adjusting the transferred proportion r. However, the pattern that the hierarchy structure follows to change is based on Eq. (8). We hope that once we reach a^*, we can choose some specific transferred proportion to make a^* stay the same. Therefore, we can determine that in a specific period, which group of students should the educators focus on and what should they do. In other words, we will distribute the workload based on the situation of the students and the levels they are at to maximize the working efficiency.

We build a model

$$\begin{cases} \min_{r} \ D(a(1),a^*) \\ s.t. \ a(1) = a(0)(Q+w^T r), \\ r_i \geq 0, \ \sum_{i=1}^{k} r_i = 1 \end{cases} \tag{13}$$

We perform iteration on the model and have the adjustment table of the transferred proportion r as follow (Table 1).

Table 1. The adjustment table of the transferred proportion

T	1	2	3	4	5	6	7	8
r(t)	0.32	0.33	0.367	0.403	0.525	0.649	0.754	0.952
	0.32	0.35	0.413	0.426	0.375	0.303	0.213	0.043
	0.24	0.243	0.17	0.13	0.056	0.034	0.023	0.012
	0.12	0.077	0.05	0.041	0.034	0.014	0.010	0.003
a(t)	0.1	0.112	0.144	0.167	0.192	0.213	0.224	0.234
	0.1	0.112	0.144	0.167	0.192	0.213	0.224	0.234
	0.3	0.287	0.242	0.223	0.189	0.173	0.167	0.172
	0.5	0.489	0.47	0.443	0.427	0.401	0.385	0.360

The variation trend of r(t) shows the population proportion that transferred to some hierarchy per year. The educators could adjust their workload that distributed to different levels according to the transferred proportion in order to maximize the working efficiency. The values of a(t) describe the developing trend of the population proportion of every level. We can keep the system close to its ideal condition by adjust r, the transferred proportion, to time to keep the system with stable structure. Therefore, we can keep the sustainable development of the system structure.

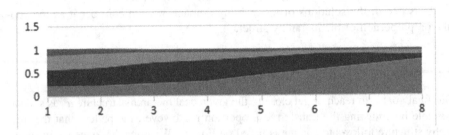

Fig. 4. The variation trend of the educators' work emphasis (Color figure online)

In Figs. 4 and 5, the regions with different color denote different interaction parts, where, blue denotes trust, orange indicates agreement, gray expresses participation and yellow is for development.

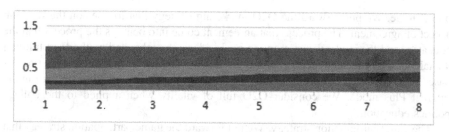

Fig. 5. The development trend of the population proportion at every level (Color figure online)

5 Exploration on the Strategies and Methods of the College Politics Education in the Internet Plus Mode

We have the main characteristics of the young students who are deeply influenced by Internet and the basic patterns of the college politics education in the Internet Plus mode. Based on that, we supplement exploration on the college politics education to discover the best strategies and methods to transfer the tentative idea into practical work.

5.1 The Basic Strategies

The conversational cooperation strategy. We put forward the strategy based on the require of the changing subjects of the college politics education in the Internet Plus mode. The conventional mode that consists of subjects and objects becomes the mode that consists of the educators and the students as two subjects. Therefore, we need take the conversational cooperation strategy. We take the inner needs for participation into consideration when we make the strategy. In addition, the strategy contains the require for pluralities, possibilities and playfulness of the students who are the subjects in the educational activities. It is the important guarantee for motivating the basic working aspects including trust, agreement, participation and development.

The service driven strategy. As for the working aspect of trust, we put forward the service driven strategy. If we want the students to trust the educators, the educators must trust the students first. The educators need to provide the students with more services for their development and growth. Beside that, they need use Internet technology to improve their service and make their service more adequate, more efficient, more accurate and more personalized. Based on the Maslow's hierarchy of needs, the educators could carry out their work in the order that from the low hierarchy of needs to the high hierarchy of needs. They could actively blend in with the students and listen to their require, feel their growth, discover the main problems. Based on that, they could adjust the directions and methods of designing their service. Therefore, they could satisfy the students' needs from the low hierarchy to the high hierarchy. In that case, the students will have strong feel of security for the educators and to some certain extent depend on them. Therefore, the educators and the students trust each other.

The O2O marketing strategy. We use O2O to represent Online to Offline, which combines the offline business with the Internet and makes Internet the platform of offline trade. We put forward the O2O marketing strategy that focuses on the working aspect of agreement. The process that agreement come into being is the process that the educators and the students agree on the goals, ideas, thoughts and methods of training the talents. The educators need to use different ways and methods to realize the process based on trust. We could promote the process using the O2O marketing strategy in the Internet Plus mode. We consider O2O full of vitality when applied to the college politics education.

The game participation strategy. We put forward the game participation strategy that focuses on the basic work aspect of participation. We could fully exert the latent energy of the students' characteristic of entertainment. We know that the educators fully research on and well design their educational carrier and carry out practical education activities. If we want the students to participate in them actively, we should consider the process of the students participating in the educational practice as the process of the students participating in the game. We could use the strategy to update the top-hierarchy design, the middle-hierarchy design and the terminal design. Therefore, we could provide the students with the experience of mastering their growth. In this case, we could make the students accept the college politics education contents imperceptibly.

The digital operation strategy. We put forward the digital operation strategy that focuses on the work aspect of development. From the conversational cooperation strategy to the service driven strategy, the O2O marketing strategy and the game participation strategy, we can see the development of the educators and the students. It is possible for us to comment on the result and perform real-time feedback on the strategies and working patterns in the Internet Plus age. We need use the digital operation strategy throughout the design of the entire process. We should perform integral top-hierarchy design, continual track, summary, analysis and data insight on the data including the data of the service design and application, the O2O marketing data, the behavior data of the educators and the students and the evaluation data on the final results. Based on that, we construct the development data reports of different hierarchies and classifications to support the comments and the decisions and to promote the educators to adjust their methods for different strategies. Therefore, we run the integral strategies into a forward loop.

5.2   Analysis on the Best Strategy Based on the AHP Method

We use the AHP method [8] to quantify the weight of every factor to illustrate the influence of Internet on the college politics education. We make educational strategies based the weights of factors.

We set the educational decisions as the target hierarchy. We have three factors at the criteria hierarchy including communicating and playing interactive games with students to carry out education work, using the PC and Wechat to carry out education work and using the network platform of the university life to extract the information of education. Then we use the five basic strategies as the alternative hierarchy including the conversational cooperation strategy, the service driven strategy, the O2O marketing strategy, the game participation strategy and the digital operation strategy. Therefore, we have the structure of the AHP method as follow (Fig. 6).

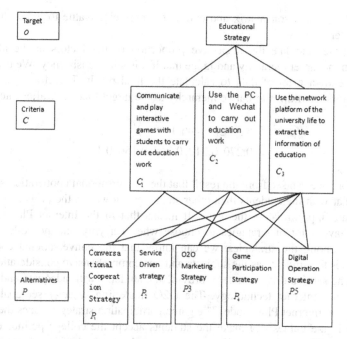

Fig. 6. The hierarchical structure of the relationship between the education and the decisions

We construct the comparison scale in Table 2 and calculate the value of the random consistency index, which is denoted as RI, in Table 3. Then we use the AHP method to calculate the weights of the factors.

Table 2. Comparison scale

Scale	1	2	3	4	5	6	7	8	9
	Same		Better		Fine		Good		Excellent

Table 3. The value of the random consistency index RI

n	2	3	4	5	6	7	8	9	10
RI	0	0.58	0.90	1.12	1.24	1.32	1.41	1.45	1.49

Firstly, we construct the comparison matrix O-C. We compare the importance of the criterion C1, C2, C3 to the target hierarchy. It means that we will compare every pair in the three factors of the criteria hierarchy. Then we have the comparison matrix and solve out the max eigenvalue of it. Therefore, we could judge that if it is with consistency.

Secondly, we compare the importance of the five educational decisions at the alternative hierarchy to the criteria hierarchy. We compare every pair of them to construct the comparison matrix and extract the max eigenvalue to judge that if it is with consistency.

Finally, we calculate the respective proportion of the factors at the alternative hierarchy in the target hierarchy and judge that if it is with consistency. We use statistic method to extract a lot of data to calculate the final result. Therefore, we have the weight vectors from the alternative hierarchy to the target hierarchy after integral sort.

$$
\begin{aligned}
w &= (w_1, w_2, w_3, w_4, w_5, w_6)^T \\
&= (0.270, 0.212, 0.247, 0.156, 0.108)^T
\end{aligned}
\tag{14}
$$

Therefore, we can see from the result that the conversational cooperation strategy is more popular with the students than other strategies. However, the proportion of other strategies are approximately the same. It means that in the Internet Plus mode, the educators have a lot of strategies to choose when carrying out the college politics education. Among the strategies, we can know that the conversational cooperation strategy takes the students as the subjects and has comprehensive consideration for the students. The service driven strategy provide the students with adequate and practical service by the Internet technology. The O2O marketing strategy set goals for the students in the Internet Plus mode. The game participation strategy makes the students participate in the games and make the students accept the college politics education imperceptibly. The digital operation strategy use the method of big data to discover new information of the students.

Appendix: Author Introduction

1. Jiawei Ren (1985), male, born in Hulin, Heilongjiang, tutor, the secretary of the youth league in School of Reserve Officers, Harbin University of Science and Technology.
2. Lina Shan (1976), female, born in Shandong, associate professor, the dean of School of Reserve Officers, Harbin University of Science and Technology.
3. Xiaohui Meng (1988), male, born in Shanxi, tutor, the secretary of the communist youth league in School of Computer Science and Technology, Harbin Institute of Technology.

References

1. BaiduPedia. http://baike.baidu.com/link?url=utus9E-OyxA296WkG6OZ3bBi60dxu4SS2SU Y1mXOIDwXGDXy_SRt40c0Imm25Yht_Py0uEPGucNks3lBdNrUEa
2. Huang, W., Ren, J.: The content and mode of moral education management in colleges and universities. Cent. Bridge **9**, 92–93 (2012)

3. Craft, A., Dai, Y., Shen, J., Zhang, H.: Creative Education and Social Development Renditions: The Future of the Creativity and the Education: Learning in the Digital Age, pp. 132–137. East China Normal University Press, Shanghai (2013)
4. Meng, X., Ou, J.: Research on the psychological factors of immersive feeling formation in the game. Sci. Technol. Innov. Her. **13**, 216–217 (2011)
5. Che, H., Gu, J.: The scale-free networks and its scientific significance. Syst. Eng. Theor. Pract. **24**, 82–88 (2004)
6. Shi, D.: The scale-free networks: basic theories and application. Univ. Electron. Sci. Technol. J. (2010)
7. Zhai, X.: The nature of trust and its culture. Society **1**, 1–26 (2014)
8. Chen, D., Liu, F., Niu, B.: Mathematical Modeling, 173 pages. Science Press, Beijing (2007)

Teaching Reform and Innovation of Communication Principles Curriculum Based on O2O Mode

Aili Wang[1(✉)], Jitao Zhang[1], Bo Wang[1], Lanfei Zhao[1], and Rui Kang[2]

[1] Higher Education Key Laboratory for Measuring and Control Technology and Instrumentations of Heilongjiang, Harbin University of Science and Technology, Harbin, China
aili925@hrbust.edu.cn
[2] State Grid Harbin Electronic Supply Company, Harbin, China

Abstract. Since the communication principles content are abstract and complex, and systematic, students generally reflect this course is difficult to learn, understand and master, teachers also feel difficult to teach. Therefore, how to improve the quality of teaching of communication principles curriculum is the key problem in teaching process. The Online to Offline (O2O) hybrid teaching mode can realize classroom flip, further deepen the curriculum reform, make curriculum quality standards, recard and optimize teaching content, refine teaching requirements, improve teaching methods and the construction of curriculum resources, curriculum construction quality assurance system. It explores the reform of other courses for Communication Engineering Department, provides technical support and accumulates valuable experience for subsequent courses to complete O2O mode teaching.

Keywords: O2O · Communication principles · Teaching reform

1 Introduction

Communication principles curriculum is not only system application of professional courses based on higher mathematics, linear algebra, electronic circuit, signal and system, digital circuit, probability theory and stochastic process, but also is the basis of spread spectrum communication, communication counterwork, software radio, optical communication, satellite communication, mobile communication, communication network, intelligent transportation systems, air traffic management, mobile Ad Hoc network and other high-end professional courses, playing an important role in communication, electronic and information specialty [1, 2].

From the domestic and foreign classic textbook it can be seen, the main contents of communication principles curriculum modules is based on communication system, including communication system, analog communication system, analog signal digital transmission and digital baseband transmission system, digital modulation system, optimal receiver, synchronization, channel encoding, in which most of the chapters is related to digital communication system, and the simulation of communication system [3–5]. Some colleges and universities have prior courses of electronic circuit (one) and

W. Che et al. (Eds.): ICYCSEE 2016, Part II, CCIS 624, pp. 78–82, 2016.
DOI: 10.1007/978-981-10-2098-8_11

(two) (i.e., low frequency electronic circuit, communication circuit etc.) introduced too much, so the teaching focus on digital communication system [6–8].

At present, all colleges and universities set communication principle course for communication engineering, electronic science and technology, communications, electronics, information specialty, even many colleges and universities set related course for automatic control, computer and other professional specialty.

The Online to Offline (O2O) hybrid teaching mode can realize classroom flip, further deepen the curriculum reform, make curriculum quality standards, recard and optimize teaching content, refine teaching requirements, improve teaching methods and the construction of curriculum resources, curriculum construction quality assurance system. So the communication engineering teachers shall carry out teaching reform and innovation of communication principles based on O2O mode [9, 10].

2 Key Problems to Perform Teaching Reform

2.1 Curriculum Design

On-online teaching includes the teaching of introduction, channel and noise, the baseband digital signal transmission, digital signal transmission, frequency synchronization principle and channel encoding. Digital transmission of analog modulation system and analog signals are for offline teaching. Not simple chapters are divided learning content online or offline, but also require selection of teaching methods based on combination of characteristics chapter knowledge point, the chapters are scattered and knowledge points are put in the unit, further combing and subdivision.

2.2 The Teaching Fusion Method Combining Online and Offline Teaching

Teachers construct the online resources and students complete certain hours online learning, thus offline teaching highlight the key and difficulty, and explain the derivation and focuses on students' questions, to achieve open online and offline teaching mode and make the real learning time over previous traditional teaching hours. The teachers answer questions online raised by the students and carry out good interaction, so that teachers will continue to deepen the understanding of curriculum content and explain the knowledge points excellently.

2.3 The Construction of Online Resources

The teacher use "PPT + broadcast" pictures to strengthen teacher-student interaction. Teachers can switch between the PPT and the electronic blackboard through PPT brush teaching; Set the function of raising questions to simulate the real classroom experience; Teacher set the answer/discussion forum, timely answer to questions raised by students improving learning atmosphere; The teachers can participate in online discussions, initiate the topic, play the role of online tutoring, realize the occupation development and promote the knowledge update; The teacher gives a reply questions and the corresponding comprehensive work curriculum for in a timely manner, to enhance feedback learning in online courses, so that the teacher adjusts the teaching program according to the students to.

2.4 The Reform of Teaching Methods Offline

Teaching offline is mainly to answer and explain the problems and difficulties existing in online learning for students online. Writing on the blackboard can be more detailed and thorough, easy to derivate learning content, such as the formula derived signal-to-noise ratio, bit error rate. And drawing a single sideband (SSB) spectrum modulation and demodulation can help the students to understand the related formula, drawing the baseband waveform to understand the eye and so on. Students have sufficient time to clearly understand the relevant teaching content and courseware, making up the disadvantages that are too fast and easy to forget.

2.5 Practice Teaching Reform

Various types of experiments are fused for practice teaching including the integration verification type, design type and comprehensive type. Taking into account the different levels of students, the experimental contents are not only software programming and debugging, such as the simulation experiment of communication principle by Matlab, View System, as well as the use of FPGA hardware circuit design experiment, so that students at different levels have the opportunity to play to their strengths and expertise.

2.6 Evaluation Method Reform

It is important to combine a variety of assessment methods, focusing on process assessment. In addition to the final examination results, scores of assessment include online learning time, two integrated operations, experimental results and discussion online. Increase the proportion of the total score usually scores, encourage students to learn the key and difficult chapters online on to listen to other students, answering questions can be appropriate to be given extra points, changing from passive learning to active learning.

3 Research Methods to Realize Teaching Based on O2O

3.1 The Construction of Online Teaching Resources of High Quality and Diversification

56 h of traditional teaching hours to concentrate for 36 h online classes and 20 h offline classes, avoiding the online and offline teaching content repeated. So the teacher needs to sort out the reasonable allocation of teaching content. The use of Prezi, Camtasia Studio V6.3 and other software can achieve arbitrary switching between the PPT and the reality of teaching, the key knowledge points are shown using Matlab simulation

demo mode, enhancing the perceptual knowledge and simplify the understanding of the concept and implementation of flipping the classroom teaching. At the same time building the network platform meets the users' visit at the same time with the quality of smooth video playback. "High quality fragmented taped course" + "instant online FAQ" can truly achieve interactive teaching content, improving teaching effect of online video.

3.2   A Variety Classical Teaching Methods Fused in Offline Teaching

After the students carry out online learning resources, student learning has the most understanding, the classroom education are "well prepared" by the students, the teacher will highlight the emphasis and difficulty and answer questions raised by the students in offline teaching. Inspire students thinking and related interactive discussion, need to use the known knowledge gradually extended to new knowledge. Collaborative teaching is mainly carried out through work, group learning and grouping experiments. Job content is generally the study of specific issues, such as introducing and summarizing the frontiers of knowledge, or a question understanding and view and so on. The submitted papers and defense class are the main forms. By looking for comprehensive information, stimulate the students to interest and active participation in the teaching purpose. Students are divided into groups for group learning and group experiment in accordance with the learning situation to achieve resource sharing between students, learning objective. This kind of way requires the teachers to carefully select topic, organize, check and give a thorough and comprehensive guidance.

3.3   Strengthen the Experimental Teaching to Improve Students' Ability

Improve the synchronization of the experiment class, achieve synchronization of the experiment and the theory teaching. For example, the sampling theorem, the pulse encoding modulation experiment PCM, pulse amplitude and phase shift keying experiment and other experiments, can be arranged after completing the related theory course, so it can combine theory teaching and classroom discussion, improving the students' mastery of theoretical knowledge degree. To increase the design of experimental content, teachers layout work form and give the completion of the indicators of the experiment, students are required to make full use of existing laboratory equipment to design the plan.

The design of experimental content arrangement should reflect the flexibility and innovation. For example, design digital baseband signal transmission system using the System View software, including the verification of digital baseband transmission without distortion conditions and Nyquist sampling criteria, especially the impact of noise on the performance of the digital baseband transmission system produced by eye demo.

Increase MATLAB programming software communication system simulation. Let the students realize BPSK in the simulation of Rayleigh channel baseband communication system by MATLAB, the system consists of four documents, in addition to the main program file, three sub procedures are transmitter, channel and receiver. The main program calls this three sub functions. Let the students themselves to write such a micro software wireless communication system, communication system and enable them to understand the Rayleigh channel up to a new level, but also for their future to do scientific research foundation simulation. Based on the requirements of their simulated bit error rate with the change of SNR performance curve, BER curve and the Gauss channel are compared, this performance difference of two channel are obtained, and let the students answer the difference.

4 Conclusion

Through the implementation of teaching methods reform of communication principles course based on O2O, taking into account the needs of theory teaching, improving students' interest in learning, the experimental project and constantly adjusting the experimental requirements. First of all, improve learning interest of the communication principle course, the failure rate should be decreased; Secondly, enhance students' interest in electronic communication system design, increase the number of enrollment and the National Electronic Design Contest; Again, lay a good theoretical basis the students who regard the principle of communication as a professional course of post-graduate students and graduation thesis for the design direction of communication. For the construction of teachers, teachers work together to build a teaching model based on O2O, to improve teachers' teaching communication and the exchange of experience, improve the level of teaching at the same time.

References

1. Huang, W., Chen, L.: A new era of mobile Internet leading O2O. Commun. Enterp. Manag. **9**, 77–79 (2013)
2. Liang, J., Liu, B.: Research on O2O teaching mode from the perspective of synergy theory. Vocat. Tech. Educ. **11**, 33–36 (2015)
3. Wang, R., Du, C., Zhang, B.: Research on the teaching mode of O2O based on mobile social network. China Audio Visual Educ. **12**, 113–119 (2015)
4. Ding, Y.: Research on the O2O teaching mode and management mechanism based on flipped classroom concept. Univ. Educ. Manag. **1**, 111–115 (2016)
5. Yu, D., Jun, W.: A brief discussion on the teaching mode of O2O course. Talent **36**, 212 (2014)
6. Gu, S.: Research on the application of interactive teaching method in the teaching of college courses under the online and offline (O2O). Teacher **17**, 89–90 (2014)
7. Gao, D.: Construction of modern distance education O2O learning support service system. Educ. Mater. **23**, 153–154 (2014)
8. Yang, L., Liu, Y.: The application of interactive teaching method in the course of the online and offline (O2O) teaching. New Curriculum Res. Mag. **11**, 96–97 (2015)
9. Xi, C.: The reform of the object oriented curriculum model based on online education O2O. Comput. Knowl. Technol. **13**, 115–116 (2015)
10. Yu, Y.: Research on the design and application of O2O flip classroom based on Mobile Learning. China Audio Visual Educ. **10**, 47–52 (2015)

The Research of Excellent Talent Training Model Reform and Practice Innovation Aimed at Computer Specialty

AChuan Wang[✉], Chang Hou, and RuiGai Li

College of Information and Computer Engineering, Northeast Forestry University, Harbin, China
wangca1964@126.com

Abstract. Combined with "Excellent engineer education and training project" of the instructions and general requirements that the ministry of education presents, The paper analyses the problem of the existing computer specialty talent training project, according to relevant requirements and implementation issues of "The Excellent engineer" program, combined with running situation of our school, explored computer science implementation methods and specific measures of "The Excellent engineer" training project, and analyzed the talent training quality and service ability. Practice shows that education "Excellent engineer education and training program" in our school had achieved good effect, for excellence engineers engineering practice education quality improvement has a certain reference significance.

Keywords: Computer specialty · Excellence engineer · Talent training mode

After entering the stage of popularization of higher education, our country's economic and social development put forward new requirements for Excellent engineering talents and the cultivation of innovative talents. "Excellent engineers training project" is earnestly implementing important reforms of 《 the national medium and long-term education reform and development plan outline (2010–2020) and 《 the national medium and long-term talent development outline (2010–2020), is also the important steps made by the engineering education in China country towards engineering education power, to promote higher education faces the society needs to cultivate talents, to engineering education of the personnel training quality having great importance comprehensive demonstration guide role and far-reaching significance. "Excellence engineers program" is aimed at promoting the reform of higher engineering education, cultivating social needs of all kinds of engineering and technical personnel [1, 2].

Computer specialty is a mainstream of the current information, in the engineering professional enrollment is also the most of the largest professional social demand. It has the characteristic of knowledge update quickly, high mathematics theory knowledge requirement, strong operability, strong practicality. At the same time, Enterprises for the computer professionals is very tall to the requirement of practical ability, innovation ability and engineering ability [3], therefore, the computer professional talent training is an important project of developing a target of excellent engineer. At present, how to build a computer professional personnel training mode aimed at "The Excellent project", optimizing the teaching system and the implementation plan is our current problem to

W. Che et al. (Eds.): ICYCSEE 2016, Part II, CCIS 624, pp. 83–91, 2016.
DOI: 10.1007/978-981-10-2098-8_12

be solved. In order to adapt to the training objectives and requirements of "The Excellent engineers project", according to our university computer professional education present situation and the running situation, combined with their own characteristics, reform and innovation excellence engineer talent cultivation system, explored our own characteristics of the "Excellent engineer" personnel training mode reform, to cultivate computer class excellent engineers talents to satisfy the needs of the new era is having important theoretical significance and practical significance.

1   Existing Deficiency in the Current Computer Professional Talent Training Scheme

1.1   The Contradictions Between the Single Training Mode and the Demand of Diversity Talent

At present, some schools single, traditional, rigid, backward training mode has not fundamentally changed, and has become a serious constraint main problems of improving education quality in universities. To the problem of higher education is still exist: (1) The contradictions between the single training mode and the demand of diversity talent; (2) Oneness of talent training mode and the contradiction between diversity of institutions of higher learning; (3) Oneness of talent training mode and the contradiction between diversity of the characteristics of students. Therefore, to solve the oneness of the contradiction between and social demand diversity, the contradictions characteristics in students diversity, make out the excellent talents to meet the needs of national economic development of talent, must not only promote the personnel training mode reform, and is imminent.

1.2   The Talent Training Scheme Too Much Emphasis on Theory Course

Some university personnel training plan formulation mainly adopts to public basic course, professional basic course and specialized course curriculum pattern, therefore, professional required courses, other course structure, and the practice teaching is set to be optimized. Curriculum system and the adjustment of industrial structure is not adapt, course content is based on the theoretical knowledge as the main body, obsolete teaching content and course content lags behind the updating and development of professional technology; Still use a teacher-centered teaching, limited to textbook knowledge teaching, emphasis on theoretical knowledge. So go down to bondage and suppresses the students' innovation ability and innovation spirit, students' learning enthusiasm not effectively stimulate independently, turned down the opportunity to explore issues. Therefore, cultivating students basic can't satisfy the computer professional talent training plan and requirements of outstanding engineers "innovation".

1.3   Not Enough Emphasis on Practice Teaching Link

In some course practice teaching process, still stays in the non-computer majors, students passive imitation method of teaching. And most of the experiment is given priority to with the verification experiment, comprehensive experiment design and less, the lack of independent experiment courses, class hour is enough. Course practice teaching link between cohesion is not quite close together, only to stay in consolidating theory knowledge stage. Make the students' ability to analysis and solve practical problems, design ability, innovation ability, comprehensive application ability is not fully improved. On the other hand, the school did not establish fixed practice teaching base, enterprises to provide internship students internship is less; Insufficient recognition of the school to practice at the same time, lax, lead to the campus internship perfunctory, practice outside for a walk form; Course design of east copy west spell to cope with the teacher, some students even in the graduation design to deal with online purchase. The result must be graduates lack experience to solve practical problems and the ability of analyzing and resolving problems, which is difficult to competent for his job.

1.4   The Depth of the Enterprises to Participate in the Talent Training Scheme

School joint enterprise, the common participation in the process of talent training, beneficial to cultivate talents of engineering quality, is one of the "Excellent project" to actively promote personnel training mode. Traditional classroom education as the main body of education mode, cultivate the students' general lack of engineering practice ability and innovation ability, and nowadays the social and economic development and enterprise demand does not adapt, difficult to meet the needs of the social development. Some colleges and universities to carry out joint between colleges cultivate mainly: on the one hand, training employees for the enterprise, as the enterprise to provide talent pool; On the other hand, the enterprise internships for students, practice base and so on. School also less inviting enterprise and unit of choose and employ persons to participate in the talent training scheme of making process, the talent training scheme of colleges and universities to develop, it is difficult to meet the requirements of enterprise knowledge system.

1.5   Lack of "Double Type" Teachers

"Double type" teachers is an important link in improving teaching quality, but also the base of "outstanding engineers" talent cultivation and the key, "double type" teachers should be teaching is an important part of cultivating creative personnel in colleges and universities. To cultivate creative ability to meet the demand of society computer "outstanding engineers" talent must have a strong teaching staff. At present, the colleges and universities engaged in theoretical teaching occupies a large proportion of professional teachers, they are busy every day theory teaching, rarely have the opportunity to contact the actual project. Or the young teachers are mostly just out of school, no experience working in the enterprise, engineering design ability and management experience; Due to some schools cannot provides practical training for teachers go out training, further education and academic exchanges necessary funds and policy support, lead to can't go on a regular basis to the enterprise existing teachers learning, so teachers in the school only to find some simple project to simulate some engineering problems; At the same time schools existing less proportion of "double type" teachers. Such teachers to cultivate talents with "Excellent engineers" innovation ability is obviously.

1.6   Single of Teaching Evaluation Ways

"Excellent engineers training project" for cultivating innovation ability and engineering practice experience of engineers, this requests the computer professional talents culti-vation in the process of curriculum evaluation should attach great importance to the assessment of students' practical ability, practical ability. But mostly used in existing computer specialized curriculum teaching evaluation is given priority to with last request for the test in the exam, students, emphasis on curriculum theory knowledge, for the lack of practical ability and practice ability of the inspection. Causes students don't pay much attention to the computer professional practice skills and experience accumulation, and even sometimes test scores of students can't even basic programming. The teachers' teaching focus teaching guidance book knowledge, and neglected the practical ability, innovation ability and the cultivation of professional quality. To cultivate the students it is difficult to achieve the "great engineer" put forward "with engineering practice ability and innovation ability".

2   The Computer Professional Training Reform and Practice of "Excellent Engineer"

Founded computer major in our university, under the support of the school, after many years of hard work, the basic set up complete theory teaching system, more advanced experimental environment, adopted the "3 + 1" talent cultivation mechanism of univer-sity-enterprise cooperation and the establishment of a strong internship base, whether in the specialty construction or in the aspect of students to have achieved certain results.

In 2005, "Computer Science and Technology" become the key construction of HeiLongjiang province professional, in 2010 passed the provincial education commis-sion of our school computer provincial key professional acceptance, computer specialty and become the key construction of HeiLongjiang province in 2011. For the computer professional "Excellent talents" training the existing problems, combined with our school's situation, explore the reform of and implementation of the following methods:

2.1   Proposed and Implemented "3 + 1" Talent Training Mode of University-Enterprise Cooperation

Personnel training mode is practice talents cultivation idea, implement the pattern of the talent training scheme. In order to achieve the computer professional "excellence engi-neers" talents training target, training implementation with innovative, practical and high-quality talents in the field of engineering, we constantly research and explore the

"3 + 1" joint between colleges to cultivate the innovation of the personnel training mode. Mainly depends on schools and enterprises, the establishment of the outside school practice bases of personnel training, provides the student's undergraduate production internship at the scene of the enterprise practice opportunities, strengthen the study of the practice of the graduation design topic, and the cultivation of the students' scientific research quality and engineering practice ability.

Based on the 2005 the talent training scheme, constantly revise and improve the talent training scheme, beginning in 2008 with Neusoft group cooperation, the cooperation between higher vocational colleges "customised training and curriculum" replacement and credit exchange reform, build the knowledge, ability and quality coordinated development, pay attention to the cultivation of engineering ability construction of curriculum system. To the students' engineering ability, innovation ability, social competition ability enhancement has played a significant effect.

To update the teaching concept as the guide, to optimize the teaching content, integrated course system for the key, by means of teaching organization reform, to train students to acquire knowledge, the ability to solve the problem as the core, with diversified, incremental learning evaluation for security, to cultivate the students' engineering practice ability, for the purpose of cultivation plan of their 2009 level for the larger changes, proposed and implemented a "3 + 1" talent training mode of innovation, to achieve the diversification, individuation talents cultivation, to realize the seamless docking between colleges.

2.2 Curriculum System and Teaching Content Reform

On the basis of In the positive research, held experience seminars in corporate training of the graduate' soliciting opinions from the professional teacher recommendations, on the basis of the advanced teaching experience from both the colleges and universities, Hired by expert group composed of computer and business experts in our province to demonstration of the talent training scheme, according to the "computer science and technology professional teaching refers to appoint" request, according to the basic goal, to cultivate the excellent engineering application ability as the main line, through to the professional orientation, dividing into modules, optimizing course system and integrating teaching contents, formulating and perfecting the "control and embedded systems, network and software design" two professional direction. Through the expert full discussion and research, in the curriculum, the practice teaching, school assignment, professional quality training, the module curriculum have a more scientific and reasonable adjustment. Courses include classroom teaching, extracurricular teaching, students' development course; Combining experiment courses, projects teaching and science and technology competition, with skill training as the main line and the engineering practice, will strengthen students' ability to solve practical problems as the breakthrough point, establish and classroom teaching supplement each other and relatively independent practice teaching system. Pay attention to training students' practical ability, pay attention to quality education, innovative education and personality development, cultivating the students ability to analyze and solve practical engineering problems. Realizing the seamless connection between colleges of "3 + 1" training mode innovation, cultivating the outstanding student conforming to the requirements of the industry of "zero adapting period". The main reforming in the following:

1. Implementing the "3 + 1" innovation mode of personnel training, realizing the teaching in the university for 3 years, 4 years students can according to the needs of individual according the enterprise (training) docking or entrance examination.
2. According the training objectives of engineering application and the ability to practice, to build four levels (basic knowledge, engineering knowledge layer, the integrated using layer, innovation ability layer), nine modules (military training, engineering training, social practice, experiment teaching, course design, practice teaching, enterprise training, innovation of science and technology, graduate design) of the practical teaching system, throughout the whole process of talent training.
3. With professional teaching and engineering ability training for the purpose, established on the basis of C/C ++, curriculum group for the purpose of strengthening the program design ability, making the cultivation of the program design ability and the object-oriented program design and application throughout the whole teaching process, and established the corresponding evaluation system at the same time.
4. In view of the needed by the professional basic theory and basic engineering application ability, building a unified for the purpose of the public basic courses and professional basic course to lay solid foundation professional platform, providing strong support for professional direction module.
5. Established for "network and software design", "controlling and embedded system", to strengthen the practice ability training of the software engineering and computer engineering, in the purpose of the training courses (group) for docking with the enterprise.
6. In the school also set up the general school courses platform for adapting the social demand; Increased specialty courses in forestry application, for example, the spatial information technology and application, space database technology and so on.

2.3 Practice Bases Inside and Outside of the University

Improving the educational quality of computer professional education, to conform to the cultivating requirements of the "great engineer" is the key to strengthen the practical link, the key is to highlight the skills training, strengthen the construction of practice training. On the basis of summarizing the experience for many years, not only the professional construction of the "3 + 1" training outside of the university, mainly include the Neusoft, Harbin le Chen Technology company, the Zhongguancun Software Park, Harbin Huiye company, well meeting the requirements of practice teaching. At the same time, in order to improve the students' innovation ability and practice ability, under the support of the college, invested more than 200 thousand yuan to set up a school practice bases inside of the university, and ACM innovation training rooms, one college students innovation laboratory. Making the practice base is reasonable, wide coverage, internship content is closely integrated with the production practice, practice and internship effect is good, well satisfied the understanding of the professional practice, production practice and graduation practice requirements. In seventh semester students can selectively into learning in the practical bases inside or outside of the university, and in eighth semester do the graduation design.

2.4 The Construction of Teachers Team

To strengthen the construction of teaching staff is the key to the whole "outstanding engineers training plan". We encourage and create conditions to participate in the engineering practice for young teachers, to achieve the ultimate goal of knowledge skills complement each other, learning from each other. Adhering to the principle of combining full-time and part-time, creating a high quality and excellent capability of "double type" teachers group. Established in accordance with the professional service area, having two different directions by computer software, computer hardware across professional teaching team of teachers, to carry out the comprehensive teaching. Employing enterprise technical experts and senior managers as professor, hiring the rich practical experience of technical personnel to intern teachers for the students. At the same time supporting teachers to famous universities at home and abroad studies and assiduously study degree, improving the teachers' knowledge structure and education background, learn edge structure optimization. Since 2009, successively appointed a number of young teachers to the America and Europe and other countries as a visiting scholar in cooperation and communication. School teachers learn edge background rich, having the foreign education or experience studying teacher is increasing year by year, for the realization of the internationalization of education to lay a good foundation.

2.5 Cooperation in Running Schools Have a Certain Effect

Had hired well-known professors at home and abroad, entrepreneurs, and excellent enterprise engineering and technical personnel as part-time professors and make report to give lectures. Since 2009, having appointed five young teachers went to the United States as a visiting scholar in cooperation and study. To carry out the "948" project, the ministry of education, exchanging and international cooperation study. Bearing the TRIZ theory research and training, was carried out student exchange plan with Japan, Taiwan and other related university, conducted between joint training courses and credits replacement work.

2.6 The Main of Ability Assessment Reform

Combination of "Excellent Engineers project" computer professional training goal, maked the reform to the existing appraisal method, proposed combining various way of examination, such as the different stage, the process assessment, project appraisal way and so on. According to different computer professional course characteristics, period distribution, practice request, etc., to determine the phase test times, the form and examination method, changed the past to a piece of paper to assess students' learning methods. With innovation ability, the computer technology application ability assessment, specifically computer proficiency test scheme including, classroom practice + extracurricular practice + course design + question paper grades. This new examination method emphasizes on students' usual study manner and the process of examination, to encourage students to think problem to explore, promote students' active learning in knowledge, software development ability and the ability of the hardware design, constantly improve the students' practice ability and innovation ability.

2.7 Gradually Improve Personnel Training Quality and Service Ability

Due to the above a series of teaching reform, our university computer science made the following teaching effect.

2.7.1 The Basic Theory and Comprehensive Quality were Improved

Students basic theory is solid, great style of study, in recent 5 years, 100 national scholarship and national motivational scholarships reward, specialized in a variety of different ways to fully arouse the enthusiasm of the students one's get postgraduate, and set up tutorial system, and obtain a favourable effect, the postgraduate rate is rising year by year, In 2010, the rate reached 27.51 %. Combining with the professional education, by conducting various kinds of theme to promote education, Thanksgiving education, enterprise experts into the campus activities, the students practice and colorful style and campus cultural activities, 30 person won the outstanding student cadres in HeiLong-jiang province, HeiLongjiang province Miyoshi students, the 24th world university winter games volunteers above the provincial level awards, such as excellent one class won the title of excellent class in HeiLongjiang province.

2.7.2 The Enhance of the Innovation Spirit and Practice Ability

Because the student during the period of school take an active part in school, at the provincial level and above all kinds of competitions and awards, take an active part in school and the national college students' innovative projects, actively participate in the teacher's scientific research, technology development, etc., in the past two years the student to obtain the ministry of education and the school university student innovative experimental project funding 22, total budget of more than 200 thousand yuan; 15 students published research papers, included 2by EI, 3 utility model patents. Graduation thesis topic is the result of teachers research projects accounted for 60 %, double tutorial system accounted for 30 %, graduation thesis excellence always stay above 20 %. Students in various provincial competition and comparison of the above awards 132, participate in student number more than 200 person, nearly half of the students were offered the opportunity to exercise. With ACM/ICPC Asia division bronze, 2010 ACM national invitational bronze, ACM/ICPC collegiate programming contest in HeiLong-jiang province, the national college students' English contest C class won third prize, the national robot competition third prize, the national college students' mathematical contest the national third prize, the national college students' mathematical contest of mathematics in the mathematics class second prize at the provincial level, etc.

2.7.3   Service Economy Society Ability Increased Significantly

Serving local economic is an important content of personnel training in colleges and universities, is the necessary way to student's growth for students, is the inevitable requirement of national construction. Through teaching reform in our university computer professional one-time employment rate reached 92 % or more in recent years, the employment quality steadily improve year by year, and some students realize the high quality of employment, salary of 200 thousand yuan. Especially under the environment of the global financial crisis in 2008, the employment rate reached 95.02 %; Professional represented by digital forestry and so on a series of research results in many companies has carried on the promotion of demonstration, widely praised.

3   Conclusions

In "Excellent Engineers project", according to the problem of existing in the computer professional talent training mode, such as curriculum system, teaching content, evaluation method, etc., is not conducive to the future excellent engineer training problems are analyzed. According this, this paper combined with "Excellent engineers" training target of computer professional training mode, curriculum system, practice teaching, put forward the corresponding reform strategy, exploring a set of adaptation in the characteristics of its own excellence engineers computer talent cultivation system. Based on the research of the outstanding engineers training process, and explore suitable for the characteristics of excellent engineer training method, and then gradually enhance the level of the students' knowledge ability, Competencies and practice ability, to create a high level of outstanding talent for the state and society. For other colleges computer professional "outstanding engineers" training provides the certain reference value.

Acknowledgment.   This paper is supported by two projects:

1. Heilongjiang Province Higher Education Teaching Reform Project 《The Research and Practice of Undergraduate Computer Class "Excellent Engineers Training Project"》 (JG2013010103).

2. 《.NET Program Design》 Key Curriculum Construction Project sponsored by Northeast Forestry University in 2014.

References

1. Li, S., Teng, Y.: The mathematics course teaching reformation for the excellent engineer program. J. Changchun Inst. Technol. **14**, 120–122 (2013)
2. Liu, Y., Gao, G.: The reform of the college personnel training mode and strategy. Heilongjiang Res. Higher Educ. **1** (2011)
3. Luo, J., Zhou, S.: To renew the idea of engineering education to cultivate outstanding engineering talent. J. Changchun Inst. Technol. **9**, 111–112 (2011)

Thread Structure Prediction for MOOC Discussion Forum

Chengjie Sun[1(✉)], Shang-wen Li[2], and Lei Lin[1]

[1] School of Computer Science and Technology,
Harbin Institute of Technology, Harbin, China
{cjsun,linl}@insun.hit.edu.cn
[2] MIT Computer Science and Artificial Intelligence Laboratory,
Cambridge, MA 02139, USA
swli@mit.edu

Abstract. Discussion forums are an indispensable interactive component for Massive Open Online Courses (MOOC). However, the organization of current discussion forums is not well-designed. Trouble-shooting threads are valuable for both learners and instructors, but they are drowned out in the forums with huge amounts of threads. This work first built a labeled data set for trouble-shooting thread structure prediction by crowdsourcing and then proposed methods for trouble-shooting thread detection and thread structure prediction on the data set. The output of this work can be used to spot trouble-shooting threads and show them along with structure tags in MOOC discussion forums.

Keywords: Thread structure prediction · Crowdsourcing · Lightly supervised learning · MOOC

1 Introduction

Discussion forum is critical to MOOC because it provides interactive features for MOOC. Students are supposed to use discussion forum to shape their learning communities and peer learning environment. However, some researchers have indicated that students of MOOC are not well engaged in discussion forums [1]. It is believed that the interface and organization of current discussion forums are not well designed. For example, threads for different purposes (such as content-related questions, social activities and general questions) are juxtaposed and lack informative tags. As a consequence, it's difficult for users to find target information.

Some MOOC sites, like Udacity, have provided a way for a questioner to tag the role of the posts that replied to his questions, such as whether a post is an answer. But most of the posts are still lacking in role tags because either the tagging function isn't available or users just don't provide tags. Besides, a post may need different tags to support different functions. It would be helpful if automatic methods could be used to assign informative tags to posts in MOOC forums according to different purposes.

This work proposes solutions to automatically predict the structure within a thread for MOOC forum. The heterogeneous and diverse background of the learners in MOOC makes the contents in the forum more challenging to analyze compared to other

W. Che et al. (Eds.): ICYCSEE 2016, Part II, CCIS 624, pp. 92–101, 2016.
DOI: 10.1007/978-981-10-2098-8_13

online forums. With the structures in place, current contents in MOOC forum can be reorganized. For example, we can assign each post a semantic tag to show its role in a thread. The structure of a thread is related to its type. However, there are many thread types. We focused on trouble-shooting threads in this work to illustrate the process of automatic thread structure prediction.

Trouble-shooting threads refer to threads whose first post is asking for help. They contain the problems encountered by learners during the study process in MOOC, and posts in a trouble-shooting thread form a "learning conversation." They are valuable for both learners and instructors. For course instructors, these threads can be used to uncover learners' confusions and provide better explanations to these confusions in later instructions. Learners can find out whether a question they want to pose already exists in an established thread, along with the answer. Although trouble-shooting threads are very important, they are drowned out in the MOOC forums. No explicit tags are given to make them easy to discover.

Machine learning methods are good choices for automatic thread structure prediction. But in MOOC forum domain, there is a lack of labeled data for this task. In our work, crowdsourcing was used for labeling data instead of experts, which makes the annotation process easy to replicate and extend. Once we have the annotated data, supervised learning methods can be applied. We also propose a lightly supervised method for thread structure prediction, which can be used when there is no labeled data, a small amount of labeled data, or labeled data in another domain. The performances of the two different solutions were also compared in this work.

The remainder of this paper is organized as follows. In Sect. 2, related work is discussed. Section 3 defines the problem. Data set building process is described in Sect. 4. The proposed methods are given in Sect. 5. Section 6 shows the experiments settings and results. Section 7 concludes our work.

2   Related Work

Online forum is a rich knowledge resource that has drawn lots of interest from researchers. Forums in online education have been researched extensively even before MOOC came into being.

The users of online education forum before MOOC usually came from traditional classrooms or remote education and the number of users was about one hundred. [2] proposed a rule-based recommendation framework for a class forum with 110 registers in "Comtella Discussions platform", which can save students' time by pointing the student to relevant posts. In order to help learners to improve collaboration learning management, [3] inferred learner collaboration levels by the Expectation-Maximization clustering method with the activities of learners in forum. [4] analyzed the patterns of annual, sessional, daily and hourly user behaviors in online forums with a large-scale multi-year sample of Charles Sturt University online supported forum. [4] showed how to manage students' activities by using data mining methods to discover behavior patterns in education forums. [5] proposed a genre classification system to classify a posting as an announcement, a question, clarification, interpretation, conflict, assertion, etc. The data set came from a discussion forum of Moodle CMS used by a public senior high school in Taiwan during 2004 and 2005.

With the recent popularity of MOOC, MOOC forum has drawn a lot of researchers' attention. Forums record explicit students' activities. It is valuable for student behavior analysis and enhancement of teaching effectiveness. Currently, research on forum in MOOC mainly analyzes the forum from a macro perspective. The behaviors of learners in MOOC forums were used to evaluate the learners' engagement [6, 7] and predicate their drop off probabilities [8]. A few research efforts focus on the content analysis of posts in MOOC forum. For example, [9] defined a post classification standard for MOOC forums and annotated a data set according to the standard.

Although there is little direct research on thread structure analysis for MOOC forum, some research on thread structure analysis in other online forums are closely related to this work. [10] learned online discussion structures by a conditional random fields (CRF) method. Because only the replying structure was learned, thread types weren't considered in their work. [11, 12] learned a more complicated thread structure specially for trouble-shooting threads over a technical web forum. They assumed the trouble-shooting threads were pre-selected. [13, 14] extracted question-answer pairs from online forum threads, which could be taken as an application of thread structure prediction. Our research distinguishes itself from previous work, because we predict the thread structure after thread classification.

3   Problem Definition

The target of this work is to predict the thread structure for MOOC forum. Thread structure is related to thread types. It's necessary to know the type of a thread in order to predict its structure correctly. This work focuses on the trouble-shooting thread. This section defines the trouble-shooting type thread and thread structure prediction problem.

Formally, let $T = \{X_0, X_1, \ldots, X_n\}$ be a set of thread discussions from online forum; each thread X_n consists of individual posts $\{p_0, p_1, \ldots, p_{(m-1)}\}$ arranged in chronological order.

3.1   Trouble-Shooting Thread Definition

If the initiator post p_0 of a thread X is asking for help, then the thread X is considered as a trouble-shooting thread. This definition is very similar to "Question thread" defined in [13].

3.2   Thread Structure Prediction Definition

The target of thread structure prediction is to assign each p_i a structure tag t_i which consists of two parts: Dialogue Act (DA) class (listed in Table 1) and Link Parent (LP, Post p_i is said to be the link parent post of p_j if and only if p_j is posted later than p_i and contains an immediate follow-up discussion of p_i). LP tag is denoted by the value of the relative position between the current post and its LP. The DA classes are shown in Table 1; their detailed descriptions can be found in [12].

4 Data Set Construction

In this section, we describe how a data set is built by crowdsourcing for thread structure prediction using the threads in a MOOC course forum in edX (2013 spring course MITx 7.00x, henceforth "7.00x"). There were two stages in the whole annotation: trouble-shooting thread selection and thread structure annotation.

4.1 Trouble-Shooting Thread Selection

1000 threads (with number of replies larger than 1 and less than 10) were randomly selected from 29619 threads in the discussion forum of course 7.00x.

We designed a human intelligence task (HIT) to recruit online workers (turkers) on Amazon mechanical turk (AMT) and asked turkers to decide the intention of a forum thread. Turkers need to label whether a thread is intending to ask for help.

Table 1. Dialogue act classes

Category	Sub-category
Question	Question-question
	Question-add
	Question-correction
	Question-confirmation
Answer	Answer-answer
	Answer-add
	Answer-confirmation
	Answer-correction
	Answer-objection
Resolution	Resolution
Reproduction	Reproduction
Other	Other

Each thread was assigned to 3 turkers, and the final results were obtained by majority voting in order to minimize the effect of spammers and improve the reliability of labeling. We paid $0.01 for each thread and a total of $30 was paid for this task. 78 turkers attended this task and they completed it in 5 days.

To evaluate the quality of crowdsourcing annotation result, an expert was asked to annotate the same data. The two annotation results are shown in Table 2. The Cohen's kappa value between the two annotations is 0.812.

Table 2. Trouble-shooting thread annotation result

		Turkers	
		Yes	No
Expert	Yes	561	64
	No	26	349

4.2   Thread Structure Annotation

With the threads labeled as trouble-shooting in the previous stage, we further implemented a HIT where turkers were requested to assign the structure tag for each post in a thread (except the initial post). The structure tag of a post consists of two parts as defined in Sect. 3.2: Link Parent label and Dialogue Act label (one of the 12 sub-category labels in Table 1).

561 trouble-shooting threads (agreed by Expert and Turker in Table 2) were chosen for this annotation stage. There are 1977 posts and the average number of posts per thread is 3.5. The average number of words per post is 42.

In this task, we paid $0.05 for each thread and each thread was assigned to 5 turkers. $140.25 was paid. A total of 166 turkers were involved in this task, and 125 of them had some familiarity with courses material.

A majority voting method was used to obtain the final annotation results. The results were compared with an expert's sample annotation results (15 threads/55 posts) to calculate modified Cohen's kappa values for Link Parent label and Dialogue Act label. They were 0.76 and 0.51 respectively. This data set is called "MOOC data set" in the remainder of this paper.

5   Method

Our solution for thread structure prediction includes 2 steps: thread classification and thread structure prediction.

5.1   ME Model for Thread Classification

This step is actually a binary classification problem. The aim is to detect where a thread is a trouble-shooting thread. Maximum entropy (ME) model was used to address this problem.

5.2   Methods for Thread Structure Prediction

Because a post's role in a thread is influenced by its context or history, thread structure prediction task was formulated as a sequence labeling problem in this work. Considering the supervised learning method, the CRF model is a good choice according to previous work [10, 11].

Supervised learning works fine if we have a large number of labeled data. But the reality is that labeled data are hard to find when one is faced with a new problem or new domain. To deal with this situation, we proposed a lightly supervised machine learning method to predict the structure of a trouble-shooting thread.

Lightly supervised learning is a kind of compromise between unsupervised learning and semi-supervised learning. It can estimate the model parameters with *a priori* knowledge and unlabeled data. There are several frameworks that can utilize *a priori* knowledge to do model parameter estimation. We used the Generalized Expectation (GE) criteria framework, which was proposed by McCallum [15] and is suitable for combination with discriminative model.

In practice, GE criteria were used as a term in the object function to involve the feature constraints (*a priori* knowledge) into model parameter estimation. Different score functions could be defined to express the model preferences on some features. For example, Formula 1 defined a KL divergence function to calculate the differences between the prior distribution $\tilde{\Phi}$ and model distribution $E_{(p(y_U|x;\theta))}[\Phi(x,y_U)]$ of feature $\Phi(x,y_U)$.

$$S(E_{(p(y_U|x;\theta))}[\Phi(x, y_U)]) = -D_{KL}(\tilde{\Phi}||E_{(p(y_U|x;\theta))}[\Phi(x, y_U)]) \qquad (1)$$

The feature constraints in the GE criterion could be obtained in the following manner: assigned by domain experts; calculated from feature annotation data; calculated from sample annotation data.

The GE criterion needs to be combined with the concrete machine learning model to estimate the model parameters. So a method combining CRF and GE criteria (GE-CRF) for thread structure prediction was proposed in this work. The object function of the proposed GE-CRF is defined as formula 2.

$$O(\theta) = \log p(y_L|x; \theta) + S(E_{(p(y_U|x;\theta))}[\Phi(x, y_U)]) + \log p(\theta) \qquad (2)$$

In formula 2, θ represents the parameters of the CRF model; $\log p(\theta)$ is the regularization term to constrain the size of θ; $\log p(y_L|x; \theta)$ is used to calculate the likelihood of labeled samples. It can be removed if there is no labeled sample. The Mallet toolkit was used to implement the proposed method.

5.3 Feature Description

Features used for trouble-shooting thread classification were borrowed from [13], including: number of question marks; number of question words (5W1H); N-gram features (1-g to 5-g); authorship and number of posts in the current thread.

Features used for thread structure prediction are drawn largely from the work of [11, 12]. Three categories of features were involved: structural features, semantic features and author features. The detailed feature descriptions are shown in Table 3.

Table 3. Features for thread structure prediction

Feature category	Feature name	Feature description
Structure features	Initiator	Whether the author of current post is the initiator of the thread
	Position	The position of current thread
Semantic features	qmark	# of question marks in a post
	emark	# of exclamation marks in a post
	url	# of URLs in a post
	PostSim	The relative position of the most similarity post
Author feature	PageRank	PageRank value of the author of current post

6 Experiment

This section reports our experimental results for the two steps mentioned in Sect. 5.

6.1 Thread Classification Results

With the ME model, the data set annotated in Sect. 4.1 and the features mentioned in Sect. 5.3, 10-fold cross validation average results for trouble-shooting threads are shown in Table 4. The results are comparable with results in [13].

Table 4. Trouble-shooting thread classification results

Classes	Precision	Recall	F-measure
Trouble-shooting	0.789	0.955	0.864
Non trouble-shooting	0.842	0.485	0.615
All	0.799	0.799	0.799

6.2 Thread Structure Prediction Experiment

In this part, we compared the prediction performances between supervised CRF and GE-CRF with different setting as shown in Table 5. Accuracy was used as our evaluation metric. In Table 5, all results were average over 10-fold cross validation on the MOOC data set except the results in row 2 and row 4. CNET was a data set built by [12] in another domain which had the same tag set as the MOOC data set.

The position-based baseline method proposed by [11] achieved an accuracy of 0.47. It classified all the first posts of a thread as "0 + Question-question" and all the second posts of a thread as "1 + Answer-answer".

We can see that supervised CRF trained on the MOOC data set obtained the best performance. The lower accuracy of CRF trained on the CNET data set indicates that the label distribution differs between the two data sets. The results in row 5 confirm this observation. Figure 1 shows the dialogue act category distribution differences between the MOOC data set and the CNET data set.

For GE-CRF, the method used to obtain the feature constraints is vital to the final performance. Here we present 3 ways to obtain the feature constraints and compare their performances: (1) Obtain feature constraints from an existing data set in another domain (the CNET data set was used); (2) Obtain feature constraints by expert assignments; (3) Obtain feature constrains from labeled MOOC data. The third way (row 7) realized the highest accuracy in all GE-CRF setting. So the feature constraints calculated from the data with identical distribution as the test set are most effective.

The score function of GE also affected the results. In Table 5, KL denoted the KL divergence score function and L2 denoted the squared difference function. KL's performance was better than L2's. But for KL, every label needs to be assigned a constraint value, which is not convenient when the feature constraints are assigned by experts.

Table 5. Thread structure prediction results

Method	Training data	Accuracy
Position-based		0.47
CRF	MOOC	0.576
CRF	CNET	0.521
GE-CRF	Feature constraints calculated from labeled CNET data + unlabeled MOOC data	0.423 (L2) 0.461 (KL)
GE-CRF	Feature constraints assigned by an expert + unlabeled MOOC data	0.495 (L2)
GE-CRF	Feature constraints calculated from labeled MOOC data + unlabeled MOOC data	0.501 (L2) 0.517 (KL)

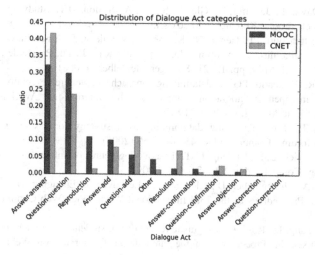

Fig. 1. Dialogue act category distributions of MOOC data set and CNET data set (Color figure online)

The thread structure prediction performance of this work is lower than what was achieved in CNET data set in [11]. The reason for that may be 2-fold: (1) Interactions in MOOC forum threads are more diverse than in CNET forum. The performance of a position-based baseline could be a kind of evidence: in the CENT data set it was 0.515, while it was 0.47 in the MOOC data set. (2) The annotation consistency of the CNET data was higher than that of the MOOC data because the kappa values of the CNET data (LP: 0.78, DA: 0.59) are higher than those of the MOOC data.

7 Conclusion

This work defined the trouble-shooting thread selection problem and thread structure prediction problem for MOOC forums. ME model was used to address the trouble-shooting thread selection problem and CRF and GE-CRF were adapted for thread structure prediction problem. The contributions of this paper include: First, We built an annotated data set by crowdsourcing for understanding the interaction of trouble-shooting threads in MOOC forums. Our practice showed that crowdsourcing is a cost effective way to annotate forum data. Second, we proposed a framework for thread structure analysis from scratch, which includes two steps: thread classification and structure prediction. Third, we provided supervised and lightly supervised methods for thread structure prediction in different situations and compared their performances.

Acknowledgment. This work is sponsored by Quanta Computers, Inc. under the Qmulus Project and National Natural Science Foundation of China (61572151 and 71573065).

References

1. Lori, B., David, P., Jennifer, D., Glenda, S., Ho, A., Seaton, D.T.: Studying learning in the worldwide classroom: Research into edx's first mooc. Res. Pract. Assess. **8**, 13–25 (2013)
2. Abel, F., Bittencourt, I.I., Henze, N., Krause, D., Vassileva, J.: A rule-based recommender system for online discussion forums. In: Nejdl, W., Kay, J., Pu, P., Herder, E. (eds.) AH 2008. LNCS, vol. 5149, pp. 12–21. Springer, Heidelberg (2008)
3. Anaya, A.R., Boticario, J.G.: A data mining approach to reveal representative collaboration indicators in open collaboration frameworks. In: International Working Group on Educational Data Mining, pp. 210–219 (2009)
4. Dringus, L.P., Ellis, T.: Using data mining as a strategy for assessing asynchronous discussion forums. Comput. Educ. **45**, 141–160 (2005)
5. Lin, F.-R., Hsieh, L.-S., Chuang, F.-T.: Discovering genres of online discussion threads via text mining. Comput. Educ. **52**, 481–495 (2009)
6. Anderson, A., Huttenlocher, D., Kleinberg, J., Leskovec, J.: Engaging with massive online courses. In: Proceedings of the 23rd International World Wide Web Conference, pp. 687–698 (2014)
7. Wen, M., Yang, D., Rosé, C.: Linguistic reflections of student engagement in massive open online courses. In: Proceedings of the International Conference on Weblogs and Social Media (2014)

8. Ramesh, A., Goldwasser, D.: Modeling learner engagement in MOOCs using probabilistic soft logic. In: NIPS Workshop on Data Driven Education, pp. 1–7 (2013)
9. Stump, G.S., Deboer, J., Whittinghill, J., Breslow, L.: Development of a framework to classify MOOC discussion forum posts : methodology and challenges. In: NIPS Workshop on Data Driven Education, pp. 1–20 (2013)
10. Wang, H., Wang, C., Zhai, C., Han, J.: Learning online discussion structures by conditional random fields. In: Proceedings of the 34th International ACM SIGIR Conference on Research and Development in Information Retrieval, pp. 435–444 (2011)
11. Wang, L., Lui, M., Kim, S.N., Nivre, J., Baldwin, T.: Predicting thread discourse structure over technical web forums. In: Proceedings of the 2011 Conference on Empirical Methods in Natural Language Processing, pp. 13–25 (2011)
12. Kim, S., Wang, L., Baldwin, T.: Tagging and linking web forum posts. In: Proceedings of the Fourteenth Conference on Computational Natural Language Learning, pp. 192–202 (2010)
13. Hong, L., Davison, B.D.: A classification-based approach to question answering in discussion boards. In: Proceedings of the 32nd International ACM SIGIR Conference on Research and Development in Information Retrieval, SIGIR 2009, pp. 171–178. ACM Press, New York (2009)
14. Cong, G., Wang, L., Lin, C.-Y., Song, Y.-I., Sun, Y.: Finding question-answer pairs from online forums. In: Proceedings of the 31st Annual International ACM SIGIR Conference on Research and Development in Information Retrieval, pp. 467–474. ACM Press, New York (2008)
15. McCallum, A., Mann, G., Druck, G.: Generalized expectation criteria (2007)

Training Mode of Personnel Majoring in Network Engineering Based on Three Main Lines

Hongzhuo Qi[✉], Guanglu Sun, and Zhiyong Luo

School of Computer Science and Technology,
Harbin University of Science and Technology, Harbin 150080, Heilongjiang, China
16909487@qq.com

Abstract. Network engineering is an emerging major across many disciplines, so there is yet to be an established personnel training program for this field. Moreover, there is a lack of reference material concerning personnel training. Therefore, it is especially urgent to explore a model suitable for personnel training in network engineering and study relevant personnel training systems. In this paper, the present situation of network engineering personnel training in China is analyzed, and basic orientation and training targets of network engineering major are explored. We find that network engineering personnel training is based on three main combinations: theory and practice, learning and innovation under the social computing environment, and classroom and employment experience, thereby providing constructive experience for training high-quality personnel specializing in network engineering.

Keywords: Network engineering · Personnel training · Social computing · Teaching practice

1 Introduction

Computer technology, network technology and communication technology have become masters of information society and important foundations of developing knowledge economy. Meanwhile, the three technologies have further development in the major of network engineering. The newly emerging industry has struck much attention from the various walks of life. Therefore, training of professional personnel and characteristic personnel in the aspect of network engineering has become the opportunity and challenge faced by colleges and universities.

Since Ministry of Education additionally has established the major of network engineering for the first time in 2001, nearly more than 300 colleges and universities have set up the major of network engineering in China according to incomplete statistics. Network engineering was a branch of computer science in the past. It belongs to the major of computer together with computer science and technology, and software engineering. Computer science and technology teaching steering committee established under Ministry of Education has not published unified teaching specification, knowledge system and curriculum standards aiming at the major of network engineering. There is no unified standard of network engineering in the aspects of training objects and

© Springer Science+Business Media Singapore 2016
W. Che et al. (Eds.): ICYCSEE 2016, Part II, CCIS 624, pp. 102–109, 2016.
DOI: 10.1007/978-981-10-2098-8_14

knowledge system's construction as an newly emerging major. Personnel training program of network engineering is not mature and perfect [1].

Personnel training mode and curriculum of network engineering are studied in some domestic colleges and universities. The results show that curriculum of all universities is closely related to the powerful major in the universities. For example, Beijing University of Posts and Telecommunications and Nanjing University of Posts and Telecommunications have regarded training professional personnel of computer communication engineering and communication software engineers as targets. Communication principle, modern switching principle, signal and system, modern communication network, communication software design and other courses of communication major are regarded as their professional courses. Sun Yat-sen University, Chongqing University and PLA University of Science and Technology have regarded training of personnel for computer network engineering management, design, development and network system management as purposes. Network management, network planning and design, LAN technology and network engineering, computer network security and embedded system are regarded as professional courses in the major [2, 3]. Because the universities are different in the aspects of level, industry, teaching resource allocation and education concept, personnel training mode and curriculum of network engineering are different and flourishing in different universities.

In addition, there are various problems between personnel training program and corresponding curriculum system of network engineering major. Firstly, the major positioning is backward, and the curriculum can not meet the demand of personnel. When the China's tenth five year plan is implemented, the positioning of the network engineering has focused on construction of networking engineering as well as management and maintenance of network system. Currently, the state has entered into the twelfth five year plan, how to exert the information technology's function, protect the information security and IOT are critical for development. However, the major of network engineering is weak in training students' main abilities in network technology application and network technology services. Personnel training target and specifications are slightly dislocated with social demand. Secondly, current curriculum system is cooperated with industry enterprises, but there is still some distance from real work's situation. Organizational logic of course content has low correlation with the work logic of working contents in their posts. Students may be still lack of certain adaptation ability to their own posts after graduation. Thirdly, the existing curriculum content still has certain traces of discipline system more or less, and teachers still have paid more attention to knowledge teaching in daily teaching. It will result in students' learning-weariness on one hand, and it is not beneficial for training high-skilled personnel on the other hand. Fourthly, teachers are affected by occupational habit in teaching, and they just focus on knowledge, skills and similar specific contents in the courses. They always believe that students can achieve ability if they obtain the knowledge and skills. Therefore teachers do not pay attention to curriculum structure design; however, the curriculum structure is the most important variable formation of students' vocational ability actually. Knowledge organization mode is always more important than knowledge itself. Therefore, the exploration of training mode of personnel majoring in network engineering suitable for China information society's construction demand has urgent and important practical significance to standardize the major of network engineering teaching and improve the teaching quality.

2   Personnel Training Target of Network Engineering

Personnel training refers to scientific composition of a structure composed of three major factors, personnel's knowledge, ability and quality. It should be consistent with the demand of social development and economic construction, follow up with the education teaching rules, which is consistent with the school type's orientation and personnel training target positioning. The academic standard of undergraduate education are clearly regulated in China's Higher Education law. The major of network engineering belongs to undergraduate education. The contents should depend on basic the requirements and corresponding type of undergraduate education. 'Foundation' is strengthened and 'long-term goal' is emphasized in undergraduate education. It is required that students must master basic theories and basic skills necessary for their own discipline and major systematically. They should have preliminary ability of engaging in practical work and research work of their own major [4]. Network technology is a new technology of combining computer technology and communication technology, and it is also a cross discipline of computer technology and communication technology, therefore the training targets of network engineering is different from the training targets of computer's undergraduate course major, and it is also different from the training targets of communication engineering major.

In summary, training targets of network engineering major at undergraduate level are determined as follows: to train practical personnel of network technology who can follow the socialist road; to adapt to social market economy, science and technology as well as social development needs; to develop moral, intellectual, physical, aesthetics and labor aspects comprehensively; to have solid foundation knowledge in their own major; to master basic theories of computer network and network engineering's practical technology systematically, to have innovative consciousness, practical ability, teamwork spirit and higher comprehensive quality. They are mainly engaged in planning and design, demand analysis, application development, implementation deployment, security management, operation maintenance etc. of network engineering and network system. They are competent for the post of network engineer, network architect, network test engineer, network sales engineer, etc.

3   Personnel Training Mode Guided by the Three Main Lines

3.1   Combining Theory with Practice in Order to Improve Teaching Quality

Training target of personnel majoring in network engineering is positioned to 'application and innovation' according to the positioning and requirement of national science and technology development strategy for training personnel majoring in network engineering and strong practical characteristics of network engineering major. Application and innovation are mainly manifested in teaching students how to apply basic theories and principles of network engineering to engineering practice, wherein theory learning becomes the pioneer for students to master network understanding, network design optimization and network management. Two platforms can be built in the major in order to provide students with solid basic knowledge and professional knowledge: one is a public basic course platform, which includes humanities and social science foundation courses, science and engineering foundation courses, computer basis and application courses; the other is a professional basic course platform, which includes network design planning and deployment courses, network management courses, network application development courses and professional characteristic courses, which can adapt to basic demand of the society in training applied and innovative technical personnel majoring in network engineering [5]. It is required that theory teaching system of network engineering major should be optimized, ensuring optimal and systematic teaching contents such as basic knowledge, basic theories, basic principles, etc. Discipline development frontier, dynamic trends, correlation with other disciplines, etc. also should be focused, and a characteristic curriculum system of network engineering teaching should be constructed. In addition, course knowledge system is not constant, which should be changed with the development of science and technology and social demand, and it is in accordance with the theory of scientific development concept.

Meanwhile, we should reject the traditional education idea of 'focusing on theory and ignoring practice' according to strong practical characteristic of network technology, paying more attention to the students' practical abilities, and students' employment ability and social adaptation abilities potentially. Firstly, investment should be increased in practice teaching, and teachers should carefully study and reform the experimental contents, methods and experimental means of experimental teaching of courses. They should increase the proportion of the experimental comprehensiveness, designing and innovativeness appropriately, and strengthen the training and cultivating the students' innovation ability. Course design of professional basic courses and specialized courses, topic selection mode and scope of graduation design, etc. should be fully combined with production, scientific research and social practice, realizing the situation that topics can be obtained from practice, and finally applied to practice actually. Secondly, universities and colleges should actively communicate and cooperate with well-known network equipment manufacturers at home and abroad for introducing and drawing lessons from their training platforms as practical teaching bases, forming a practical teaching system of cooperation inside and outside school, and resource sharing gradually, with various categories and high-level standard, in order to cultivate network engineering personnel with solid foundation theory and strong practical abilities.

3.2 Combining Learning with Innovation for Improving Students' Quality

Innovation is one of the most important features of human being, and it is an ability of creating new and suitable ideas and works, which is formed in acquired practice. Individual innovation mainly refers to knowledge application and the desire of using knowledge from the perspective of education. Innovation refers to an ability of creating original, adaptable and available ideas, solutions or insight for students majoring in network engineering. More and more graduates majoring in network engineering in applied universities and colleges will engage in working for actually solving problems of network technology with rapid development of network technology, leading to the requirements of application ability and innovation ability. Universities and colleges should implement innovation ability training in network engineering major, designing some innovation ability standard system and training plans in order to solve the problem of personnel shortage, low learning motivation of students, reducing learning interests and achievements faced by the major of network engineering, therefore the relationship among all factors for innovation ability training in education can be analyzed. It is also proposed that students are correctly guided for paying attention to innovation in the learning process through teaching activity and mutual influence among students.

Due to the development and application of social computing mode, the learning methods and the learning concept is undergoing tremendous changes. The greatest significance of social computing in education is the invisible wall between the real world and the classroom, the focus of teaching and learning is changing from teaching knowledge to developing innovative ability. It is innovative learning that social computing focused on, and it can improve the student's participation in the study and the independent learning ability. The realization of this model is mainly implemented by all kinds of social software applications, such as Email, instant messaging software, blogs, wikis, and social networking tools. Social software can be used to build social computing environment of teaching and management platform, integrate the network resources effectively, and promote the teachers' professional development. Many social software has been integrated into the learning system, as one of the fuctional modules. These social software provide a friendly learning support, extend the study space, enrich the learning resources, and develop the students' ability of information processing and cooperation. So the social computing mode can promote the socialization of education and learning adaptation.

In addition, teaching content and teaching method should be innovated in order to integrate innovation ability training in classroom teaching. On one hand, teachers should not only focus on the explanation of basic principle in classroom teaching, they but also display development rules behind the principle, associated new technique and common problems to students. After the students have comprehend principles and rules, they will not be afraid of new technology. New technology can also stimulate students' interest in learning more easily. Analysis on common problems behind the principle from the aspect of application can lay foundation for applying the theory knowledge to solve problems innovatively [6]. On the other hand, teaching methods can stimulate students' learning interest more easily, and have great significance to cultivate students' application ability and innovation spirit. For example, implementation of case teaching method should be strengthened in classroom teaching, the disadvantage of confirmatory teaching method which can only cultivate certain operating skills should be overcome, developing the students' subjective role. It can combine knowledge teaching with training of students' innovation ability organically, thereby laying solid foundation for the training of students' learning ability and innovation ability. In addition, project-centered teaching method can be implemented in some courses, knowledge points can be integrated to the implementation process of all projects. Then, setting up the project implementation plan, division and cooperation can be carried out according to the project implementation plan, and finally the project task can be completed jointly. We can train the students' ability of comprehensively applying knowledge and skills in discovering problems and solving problems as well as mutual communication ability more easily, and students' cooperation consciousness more effectively.

3.3 Combining Classroom with Employment for Realizing Training Target

'Application of knowledge in practice' is the ultimate goal for training students majoring in network engineering, and it is also a driving force of developing and strengthening the major. Major construction and personnel training should be combined with market demand in the future. Graduate quality is critical for competition's success and failure of network engineering major. It is one of the most important factors in long-term survival and development of network engineering major in all universities and colleges. It can be definitely concluded that information-based society requires more and more professional personnel mastering computer network theory and technique, including personnel mastering network planning and design, network system integration, network installation debugging, network operation maintenance, network management, network security, software development based on network platform, professional technical personnel for developing and promoting new techniques and new application of network.

Teaching resources should be fully organized, and structure model of courses can be designed according to cohesive relation among all courses in accordance with the guiding ideas about training applied technical personnel majoring in network engineering of 'solid foundation, wide caliber, practice emphasis, good ability, much innovation and high quality' [7]. Curriculum mainly consists of four blocks: public basic courses, disciplinary basic courses, professional backbone courses and professional elective courses, wherein public basic course is arranged by the school unitedly. These courses not only belong to required basic courses, but also embody cultural environment and academic atmosphere of the school during the setup, including the following courses: college English listening, speaking, reading and writing; advanced mathematics; college computer foundation; college physics, computer program design basis; linear algebra; probability and mathematical statistics; literature search; employment guidance and humanistic quality courses. Discipline basic course is set mainly to provide students with solid computer basic knowledge, good consciousness of network engineering major, and solid foundation can be laid for long-term development of students in the future, including the following courses: introduction to network engineering; discrete mathematics; electronic technology; object-oriented programming; data structure; signal and system; database system; computer composition principle; digital communication principle; software engineering and computer network technology. Professional backbone courses mainly include the follows: network interconnection technology; network programming; network protocol; network operating system; network planning and design. Theory teaching quality should be emphasized on one hand, practical teaching matched with the course should be further focused on the other hand aiming at teaching of backbone courses, therefore students can not only achieve professional theory knowledge, but also can be provided with related practical operation ability, realizing the characteristics and advantages of applied network engineering major. Specialty direction courses are mainly set according to the social demand and ability targets. Students can select according to their own ability and interests, including network management technology direction (direction A), network security technology direction (direction B) and network application technology direction (direction C). Students' abilities can be improved through longitudinal courses, and horizontal linkage at each direction which can make students thoroughly understanding the knowledge and self expanding.

Undergraduate students majoring in network engineering should meet the following basic requirements under current social environment: they should love socialist motherland and be willing to serve socialist modernization construction; they should have solid natural science foundation, more solid basic theory of the major, professional knowledge and skilled application skills, and understand the advanced theory and development trends of their own discipline; they should have the ability to constantly update the knowledge and solve problems as well as innovation spirit; they should have good communication ability, organization and management ability, team work ability, competitive ability, good social morality and professional ethics. Adaptive course system can be established according to the demand of economic and social development on personnel, according to personnel training target, cultivating personnel conforming to social demand with stronger competitiveness.

4   Conclusion

Applied network engineering major has the main task of training personnel in applied and innovative network major, who adapt to the demand of related posts in the society. Therefore, personnel training mode of traditional computer major should not be copied completely in personnel training process. The training mode should be constantly innovated and broken through according to major characteristics and social demand. Therefore, we can cultivate qualified personnel majoring in network engineering who are popular with the society widely. We can implement a lot of activities in the aspects of exploring and reforming personnel training mode of network engineering major through exploration in recent years, and certain achievements have been obtained. It has reference significance in training practical personnel majoring in network engineering at universities and colleges. The primary employment rate is higher than 85 %, and the postgraduate qualifying examination rate is higher than 12 % among the existing graduate students. Many graduates have passed the certification exam of 'webmaster' and 'network engineer'. Students' practical ability is greatly improved. Some students have achieved certain scientific research and development ability. Students trained by our school are highly appraised in enterprises and institutions widely according to market feedback.

However, we still have some problems, which should be improved and perfected. For example, how to improve the training program, and maximally deal with the relationship between postgraduate qualifying examination and employment; how to further strengthen the cooperation between enterprises and colleges and universities, further improve students' practical ability and solve the employment problem of students; how to apply social computing technology and the concept of concrete in E-learning. In short, training of personnel majoring in network engineering is a process of constant reform and practice, which should be further explored and studied.

Acknowledgments. This work was supported by Education and Teaching Research Project of Harbin University of Science and Technology No. 220150019, Heilongjiang Provincial Education and Scientific Projects No. GBD1211026, Scientific Planning Issues of Education in Heilongjiang Province No. GBC1211062, and the Ministry of Education's Humanities and Social Science Project No. 11YJC740048.

References

1. Jiang, J., Gao, D., Yang, A., Xie, J.: Review on research status of the construction of network engineering major in China. Comput. Educ. **12**, 123–126 (2010). (in Chinese)
2. He, B., Liu, F., Huang, X.: Research on practice ability training system of applied network personnel in colleges and universities. Lab. Res. Explor. **7**, 102–105 (2012). (in Chinese)
3. Lu, B.: Exploration and research on innovative personnel training mode majoring in network engineering. China Power Educ. **25**, 119–122 (2011). (in Chinese)
4. Li, Y., Cai, Z.: Construction of 'two-main-line' teaching system for network engineering major. Comput. Educ. **23**, 134–138 (2010). (in Chinese)
5. Jin, Y., Zou, P., Wei, Y.: Exploration on teaching model of network engineering major based on project. China Univ. Teach. **18**, 122–125 (2010). (in Chinese)
6. Chen, Z., Qu, M.: Research of applied training mode of personnel majoring in network engineering. Comput. Educ. **4**, 8–11 (2015). (in Chinese)
7. Zhu, L., Geng, Z.: Making and implementation of personnel majoring in network engineering. Comput. Educ. **12**, 32–36 (2013). (in Chinese)

Virtual Simulation Experiment Teaching Platform Based on 3R-4A Computer System

Xianjun Shi, Yingtao Zhang[✉], Lijie Zhang, and Liming Wang

Harbin Institute of Technology, Harbin 150001, China
yingtao@hit.edu.cn

Abstract. Virtual simulation experiment teaching gradually becomes the trend of the computer-related education practice. Through analyzing the problems existed in current computer experiment teaching, this paper proposes the idea of building virtual simulation experiment platform based on 3R-4A computer system and clarifies the design technology, including frame, characteristic and innovation, resource sharing and management, condition protection and so on.

Keywords: Virtual simulation · 3R-4A · Experiment teaching platform · Virtual and reality co-design

1 Introduction

Virtual simulation technology is the product of the combination of simulation technology and virtual reality based on the rapid development of the multi-media, virtual reality and network communication technology. Virtual simulation experiment is the important content of information building of the higher education, and is also result of deep integration of the computer science and information technology.

Virtual simulation experiment platform based on 3R-4A computer system constructs highly-degree simulation for experiment relying on the technology of virtual reality, multi-media, human-interactive, database, network communication, students can try some experiments in virtual environments, meeting the teaching aims in the syllabus.

2 Problems Existed in Current Experiment Teaching of Computer Courses

1. Breaking through restrictions in time, space and resources, supporting students' flexible arrangement on time, and providing a solution.
2. Some modern computer experiments could hardly be done using traditional experiments, such as large-scale network, network security attack and anti-attack experiment. On the one hand it is hard to construct large-scale network meeting the experiment requirement, on the other hand, it is not permitted to carry these experiments because it will do harm to the network in reality.
3. The teaching method MOOC is in urgent need of the support of long-distant experiment platform to improve the teaching effect.

© Springer Science+Business Media Singapore 2016
W. Che et al. (Eds.): ICYCSEE 2016, Part II, CCIS 624, pp. 110–117, 2016.
DOI: 10.1007/978-981-10-2098-8_15

4. Traditional experiment environment is lack of self-innovation and individual experiments for students.

It is of great significance and urgency to build virtual simulation experiment teaching platform and break the existence problems, such as isolated lab resources, repeating buying medical equipment and managing tedious. Platform is no longer confined to the physical spaces of labs and breaks through the restrictions in time, space and resources by promoting comprehensive utilization of cloud computing, virtualization and SDN technology.

3 Architecture of Virtual Simulation Experiment Teaching Platform Based on 3R-4A Computer System

Taking virtual simulation experiment teaching platform of HIT (Harbin Institute of Technology) for example which is supported by HIT experiment teaching platform, it provides the virtual simulation experiment platform service of computer software technology, hardware technology, network technology for students in school of computer science and technology and the other related. At present, it supports about 10 courses' teaching practice and will support more than 30 courses' in the near future.

Furthermore, it will provide full support for basic research, design, comprehension, curriculum design and graduation projects, even innovation, competition (Fig. 1).

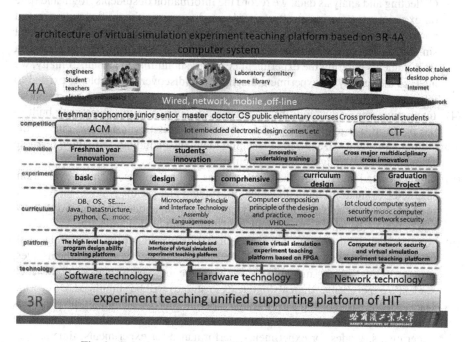

Fig. 1. Experiment teaching unified supporting platform of HIT

The feature of the computer system, 3R-4A, is providing virtual simulation experiment service for anybody, at anytime, anywhere using any methods.

4   Functions of Virtual Simulation Experiment Teaching Platform

Taking virtual simulation experiment teaching platform of HIT for example, it is made up of the comprehensive service website of the experiment teaching center, virtual simulation experiment teaching platform for computer networks and security, virtual simulation experiment teaching platform for microcomputer principle and interface, virtual simulation experiment teaching platform for remote connection based on FPGA(Field Programmable Gate Array), platform for advanced language program design training, the system of platform management and maintenance. It has these functions as follows:

(1) Information distribution: it publishes relevant experiment information of different courses, including: name of experiment course, instructor, task, experiment requirement and so on, making teachers and students master the experimental schedule and progress.

(2) Users and permissions management: according to the different roles including different class of students, instructors, managerial personnel, statistical analysis of personnel and so on, we allocate different rights to them

(3) Collecting and analysis data: we record the information of students' login and time on experiment, experiment effect, grades and so on via the unified entrance and control, uniting different sub platforms. Furthermore, by combining the virtual simulation experiment with the traditional experiment, we collect the complete data from students entering the experiment center to submitting the results. Finally, we analyze the quality of experiment teaching and discover the problems existed to improve experiment content and check mechanism by big data analysis system.

(4) Interactive communication: through forum, discussion group, guestbook built in the platform, we realize the students exchange activities inside and outside the community and unitize different teaching methods, including: self-study, learning from each other, studying among population.

(5) Grade evaluation: through the combinations of platform mechanical evaluation, instructor evaluation, mutual evaluation, student can get the experiment result in the first time and the platform allows you to submit more than one time. So students can optimize the whole experiments to achieve high grades gradually.

(6) Achievement exhibition: platform can automatically generate the experiment result of any student and any experiment which could be accessed by students themselves, instructors and other students. By showing the best performance of the each student's experiments, students' interests in studies will be stimulated.

(7) Data analysis based on big data: through different sources, including: sub platforms of courses, varieties of courses, class of students, instructors, different contents of experiments, grades for experiments, performance for experiments, time assignment in experiments, status of interaction, forum posts, we analysis these information comprehensively using big data system. We can further know the feedback

result, optimize the contents of the experiment and adopt more effective ways for the purpose of improving the experimental teaching quality

(8) System maintenance: the platform supports database backup and restore locally and remotely, cloud backup and cloud restore. It also performs the tasks such as modules update and maintenance, handles the batch resets and set separately of the different users and authority.

5 Characteristic and Innovation of Virtual Simulation Experiment Teaching Platform

5.1 Combination Traditional Experiment and Modern Virtual Simulation Experiment Together

For computer courses teaching and experiment teaching under permitting conditions, we should try our best to do practical experiments because touching and observing experimental phenomena real things give an intuition to students and might spark students' interest. Furthermore, students can gain some insights into the problems and solve problems immediately under the supervision of the instructors step by step.

(1) Virtual experiments have some features that traditional experiments haven't, and better them on the teaching function.

(2) Virtual experiments are suitable for high cost, high consumption and large-scale or comprehensive training. Such as robots, car, UAV(unmanned aerial vehicle) and other hardware experiments.

(3) Virtual experiments are the suitable candidate for MOOC/SPOC and other large-scale online courses.

(4) Virtual experiments still need to be validated by traditional experiments or under the real environment. For example, in the experiment of remote hardware virtual experiment based on FPGA, after programming in the local simulation experiments, we still have to download the code to the real experimental platform.

5.2 Virtual Simulation Experiment Teaching Platform Based on the Feature of 3R-4A Computer System Promotes the Experiment Teaching Methods and Teaching Moods Reform

3R: Remote, Real, Rank
    4A: Anyone, Anytime, Anywhere, Any type

(1) Some projects which only used to be done in the labs, can be carried out on virtual simulation experiment teaching platform

(2) Extracurricular experiments can be done using the abundant resources and equipment provided by the modern platform, strengthening the students' practical ability.

(3) The experimental part of the online course-MOOC/SPOC can be implemented.

(4) Teachers in class can demonstrate the experiments through the internet, and introduce its corresponding guiding idea and principal experimental contents.

5.3   Support to Indigenous Innovation, Multi-disciplinary Field Innovation, Students' Competitions

Virtual simulation experiment teaching platform not only supports more than 10 courses about the basic experiment instruction, designing, comprehensive experiment, curriculum design and graduation project, but also provides support to indigenous innovation and multi-disciplinary field innovation from freshman to senior, master to doctor, to students' competitions related to computer science for students home and abroad.

Through building the virtual simulation experiment teaching platform, it completely changes traditional teaching methods which are confined to the restriction of resources, space and time. To be specific, traditional experiments must be done in fixed time, fixed place and the knowledge limited to the teachers teach in class. So, breaking through these restrictions bring a large number of benefits, students can do more experiments and have more chances to train their practical ability, which meets the need of experiments of large-scale online courses, which allows students to do experiments remotely, in other place and virtually to promote the transition from MOOC to MOOE. The platform takes one big step towards teaching methods electronic, networked and information-based in order to improve teaching effective and promote students' practical ability.

5.4   Research Results Transform to Teaching Practice, Improving Experiment Teaching Quality

Center of virtual simulation experiment teaching based on computer technology is not simply introduces or buys external software, but established in higher grade talents of computer science of HIT. Under the NSFC (Natural Science Foundation of China) and Higher Education Reform Project provided by province, we have developed virtual simulation experiment teaching platform by ourselves which can be used for public elementary courses involved software and hardware. Students can do extended experiments in their spare time after finishing experiments in class to improve students' practical ability.

5.5   Student-Centered, Designing Personal Experiment

Relying on virtual simulation experiment teaching platform, students can choose experiment projects suitable for themselves, no matter what time and place. So, students can make full use of their spare time, and do the unfinished experiments after class, promoting students' practical ability to realize the idea of designing personal experiments, which will arouse students' enthusiasm.

5.6   Make Full Use of Advantages of Professional to Reform the Computer Experimental Teaching Method, Realizing Doing Experiments in the Cloud, Learning and Exercising in the Form of Group

Computer science and technology has inherent advantages in internet, cloud computing, big data and virtual simulation. So designing and developing virtual simulation experiment teaching platform is not only the trend of our times, but also our responsibility.

To realize doing experiments in the cloud, learning and exercising in the form of group, the virtual simulation experiment teaching platform should combine the platform with the MOOC platform which has a certain scale currently. So the combined platform must support a large number of concurrent accesses, where students can talk group-based about the experiments, especially large-scale discussion across schools and it should promote self-study, learning from each other, studying among population with the function of mutual exchange, automatic ranking, statistical analysis and ranking list. Also, through different group setting, students can share their opinions, which can achieve teachers' instruction and others'. In addition, it should have the ability of data processing and analysis and arouse students' enthusiasm through ranking list.

6 Sharing Resources and Expanding Scope of Application

Platform should focus on sharing high-quality teaching and learning resources, such as National Essential Courses, the online course application of MOOC, fully expanding scope of application to our whole school and other school in our country. In addition, enriching the variety and content of experiments is also an important aspect.

Concrete actions are listed as follows:

1. Fully expanding scope of applications to other curriculums, professional and more schools in our country.
2. Strength the construction of online courses of MOOC and SPOC.
3. Promoting National Essential Courses, open courses other resources shared by curriculums.
4. Using Nationally Planned Textbook to promote carry out experiments in the platform.
5. Promoting virtual simulation experiment teaching platform through providing free technically training of the use of platform.

7 Network Management, Security and Safety of the Platform

Virtual simulation experimental platform should focus on developing construction of resources, platform, team, system and so on, form a sustained experimental teaching service and ensure an organic whole of high-quality experimental teaching resource sharing.

1. Platform should be designed to ensure the safety of the platform through strict access control mechanism and security sandbox mechanism. Platform has users and their rights management. The platform should design and assign user permissions at all levels in accordance with their post and role to ensure the authenticity and the effectiveness of user's identity. For user authentication, the platform should on the basis of experimental courses, projects, time, etc. And the platform can realize billing management.

2. The platform adopts unified network anti-virus systems, intrusion detection systems and content-based information filtering systems to realize network security, reliable and efficient operation, management and maintenance. The experiment center should be equipped with automatic alarm system which unified monitoring by the school security office. The installation of monitoring systems, punch card machines, mail servers, and file servers achieves the implementation of intelligent management. The laboratory is equipped with projectors and air conditioning equipment and has strict daily management. Interior is bright and clean, and experimental environment is elegant. The logo such as the purpose of each room and the telephone numbers of safety director is clear and intact in the laboratory. The laboratory has perfect four safety prevention measures and fire-fighting equipment. Fire escape in the laboratory is unblocked and no hinder on the route. Experimental centers and laboratories have special persons in charge of safety, periodic inspection, and work safety record. Laboratory center directors and managers sign the security liability form every year. Experimental center holds regular safety education and timely information about fire, theft notification. In guiding the experimental teaching process preach laboratory safety to students. The center has no safety accidents since established.

3. The laboratory establishes platform commissioning and daily operation and management regulations to perfect platform operation, maintenance, updating, and management norms. The laboratory examines the results of experimental courses or reports at the end of each semester and students score experimental curriculum assessment which ensure the quality of experimental teaching. The experiment center staff implement combined full-time and part-time. Theory class teacher is responsible for the corresponding experimental teaching. The teacher in charge of laboratory experiment preparation and help the speaker teachers completed the experimental teaching. Large virtual simulation display devices, computer hardware device management by designated personnel. Laboratory provisions each large equipment should have at least two or more experimental teachers who is able to skillfully operate. Experimental instructors are proficient in at least two experimental courses. Part-time teachers undertake the experimental teaching, and they subject to the consent of the experiment center, and are relatively stable. In order to ensure the professor involved in experiment teaching and reform, the school of computer science proposes the management approach of the experimental guide of teachers, which can ensure the experimental teaching faculty level, good experiment teaching results and provide strong policy guarantees to improve the overall quality of personnel training.

4. The laboratory develops the virtual simulation experiment center full-time and management norms. The platform in strict accordance with standard procedures and regulations management in daily work. And the laboratory also develops the virtual simulation experiment center of experimental teaching evaluation and feedback mechanism, which improve the experiment teaching of the closed loop management.

The construction of the virtual simulation experimental platform should adhere to the ideas of "scientific planning, sharing resources, accentuating key, improve business efficiency and sustainable development". Improving high school students' innovative spirit and practice ability for the purpose. To share high quality experiment teaching resources as the core. Construction of information experimental teaching resources as the focus. We should continue to promote experiment teaching of higher school informatization construction and experimental teaching reform and development.

Industry Track

A Classification Method of Imbalanced Data Base on PSO Algorithm

Junru Lu, Chunkai Zhang[✉], and Fengxing Shi

Hit Campus of University Town of Shenzhen, Shenzhen 518055, China
{1712028923, 767445172}@qq.com, ckzhang812@163.com

Abstract. In order to improve the performance of the resampling algorithm in imbalanced data learning and the optimal combination of the weight coefficients of the base classifiers in the ensemble classification algorithm, a new algorithm based on particle swarm optimization (PSO) is proposed. In traditional sampling method, sampling rate is often set artificially. This would make the final classification can't get a optimal solution. By using the particle swarm optimization algorithm, the sampling rate of the boundary samples and the safety samples are optimized to obtain the optimal over sampling rate self-adaptivly, the weekness of setting sampling rate of traditional method would be overcome.

Keywords: Imbalaced · Re-sampling · Sampling rate · PSO

1 Introduction

Classification problem is one of the most popular problems in the field of research and production. Among them, the classification of imbalanced samples is widely used in many kinds of work, which is widely used in the classification of the natural law and the unique characteristics. For example: Internet traffic detection, face recognition, satellite photos leak point identification detection, bioinformatics, medical decision-making, credit risk analysis [1–7] and so on.

The particularity of imbalanced sample classification problem result in that traditional classification method and evaluation strategy cannot be well applied to the problem: first, the optimization goal of statistical machine learning algorithm is to minimize the loss rate, which is the highest global accuracy [8–10]. However, most of the class and minority groups have a large amount of deviation, which leads to the traditional classifier algorithm pay more attention to majority samples and fail to recognize small samples. Second, although the total number of minority samples is much smaller than the total number of majority samples, the potential information contained in it is often more valuable. For example, in the Internet traffic analysis, traffic package is a very typical imbalanced data set, which can not be processed by the traditional machine learning classifier algorithm well, however, the situation is just the opposite, the correct classification probability will be very low. Considering some attacking activities of network traffic, mean larger byte stream to analysis work and some important properties of the minority class flow inherent which result in the minority samples should be pay more attention to [11–13]. For this reason, a big

© Springer Science+Business Media Singapore 2016
W. Che et al. (Eds.): ICYCSEE 2016, Part II, CCIS 624, pp. 121–134, 2016.
DOI: 10.1007/978-981-10-2098-8_16

damage would be came about under the traditional point where people focus more on raising the accuracy which would cause more loss than the misclassification of majority samples [14].

In this paper, we mainly study a series of optimization problems by PSO algorithm. Under the goal of maximize AUC: using particle swarm optimization algorithm to choose the best sampling rate. Finally, the particle swarm optimization algorithm is used to optimize the weight coefficients of the base classifiers in the ensemble classification algorithm AdaBoost (we name the hole process DBPS). Then a series of %ACC, F-Measure and AUC results are validated by a series of experiments on a large number of UCI data sets, which proves the effectiveness of the proposed algorithm.

DBPS performed well in experiments especially in testing some illness. So the algorithm is used for testing whether people is suffering from some cancers in WeChat platform now and performing well on testing the breast cancer.

2   Basic Concepts

2.1   SMOTE Over-Sampling Algorithm

SMOTE algorithm is a popular over-sampling technology [15–17]. It is mainly through inserting copies of the minority class samples, rather than simply sampling samples that are surrounding in the minority class samples. This technique is mainly inspired by an algorithm that has been proposed in a project that has been recognized by handwriting. The main process of SMOTE over-sampling algorithm is as follows: firstly, according to each minority class samples, we will select k other minority class samples which are near from this, then between this and other K minority samples, we will apply interpolation method. Besides, we will use a random interpolation factor and over-sampling rate to generate a part of imitated minority class samples. SMOTE interpolation process, as shown in the formula (1):

$$p_i = x + rand(0, 1) * (y_i - x) \quad i = 1, 2, \ldots, N \tag{1}$$

SMOTE algorithm mainly interpolate values among similar minority class samples (i.e. adjacent near). and thus generated copy sample is representative. Therefore, the over fitting problem can be avoided in the SMOTE algorithm, and the decision space of the minority class can also be extended better. Similarly, it can also be applied to the majority of sample space, which can reduce the decision space of the majority class.

In 2013, Lou Xiaojun proposed a new method based on clustering and taking into account the boundary sample information. Method by means of unilateral selection method determine the minority class and majority class boundary sample information, by considering the boundary of the clustering of all minority class samples, carried out in the cluster the appropriate sampling, so as to produce a replica of the minority class samples and the original sample has more similarity, can represent the parent class sample space distribution. At the same time, it can increase the definition of boundary samples by using the sample of the boundary, which can increase the definition of boundary samples.

The idea of the under sampling method is to remove some specific samples according to some rule in most class sample space. and the sample data space can be basically balanced. However, there are some disadvantages in the method of under sampling, which can easily lead to the loss of important information of some representative samples.

Mixed sampling is an algorithm that fuses over sampling and under sampling algorithm, using mixed sampling to balance the data set in the process of resampling the original data set. A large number of studies have indicated that the mixed sampling method has more advantages than the single resampling method. And line over sampling firstly and then under sampling can achieve better results. In the research of this topic, the mixed sampling is adopted.

2.2   Borderline Algorithm

In this section, we mainly describe the Borderline algorithm [18], the main function of the algorithm is to determine which samples belong to the boundary in the data samples, and the boundary samples are divided into the majority of boundary samples and the minority of boundary samples. In this algorithm, we specify the imbalanced data set to be S, each of which is composed of the feature vector and the class label of the sample. The feature vector is x, the class label is y, that is, $x = \{x_1, x_2, \ldots, x_n\}$, $y = \{Maj, Min\}$. So the data set can be expressed as:

$$S = \{(x_1, y_1)(x_2, y_2), \ldots, (x_n, y_n)\} \tag{2}$$

We use S_{maj} to represent the majority samlpes in S and S_{min} to represent a minority samples in S. The process of the Borderline algorithm is as follows:

Step 1: AS for S_{min}, we use K-nearest neighbor method to find the K nearest neighbor samples in the whole data set S. And the samples are stored in the set KNN_{smin} corresponding to each S_{min} sample.

Step 2: Classify each of the samples in S_{min} into boundary samples, noise samples and safety samples by using the following three formulas:

$$|\{(x, y)|(x, y) \in KNN_{smin} \wedge y = Maj\}| > K/2 \tag{3}$$

$$|\{(x, y)|(x, y) \in KNN_{smin} \wedge y = Maj\}| = K \tag{4}$$

$$|\{(x, y)|(x, y) \in KNN_{smin} \wedge y = Maj\}| = 0 \tag{5}$$

Those minority class samples satisfy formula (3) are boundary samples. then insert the boundary minority samples into the boundary of SBmin; those minority class samples satisfy formula (4) are noisy samples. Besides, those samples which meet the formula (5) are safe sample.

Step 3: through the (2) method, we can also get the majority of the data samples are classified according to the Borderline algorithm, for the boundary samples, we also insert these samples into a majority of the class boundary sample collection

SBmaj, the noise samples are removed directly, the security of the whole sample does not carry out any operation.

Through the above three steps, we can focus on the processing of boundary samples to increase the definition of boundary and make the classification effect improve in the next mixed sampling process. In the next section.

2.3 Particle Swarm Optimization

Particle swarm optimization (Swarm Optimization Particle, PSO) algorithm is proposed by Dr. Kennedy and Eberhart in 1995 [19–22]. PSO is a kind of evolutionary algorithm. The inspiration is source to the collective behavior of the generic nature of insects, the beasts and birds, etc. In the nature, these groups of organisms will in a way that they can understand to look for food, mates and some other things. Every member of the population will learn from the experience of their own experiences or other members of the group to change the behavior mode of individuals or groups, and complete the final group of global search. From the practical example we can understand the specific process of the PSO algorithm is as follow: suppose there are a flock of birds in a particular region aimless searching for food in the position. These birds are ignorant of the specific location of things, and also don't know which direction to fly to find food in the vicinity of the location. So, how can the birds search for the optimal route to find the food? The correct approach is to get the information from other particles closer to the food, and then move to the area where the bird is closer to the food.

In computer science, PSO algorithm is a kind of computational method, it is mainly through the iterative optimization problem and a specific fitness function to try to get the best candidate solution. "particle" in PSO algorithm represents each individual in the whole group search space. We cite the real life bird swarm search example above and PSO algorithm optimization of the specific process is as follows: Firstly, we initialize the bird's position and flight speed. This speed contains particles in the next search process to run the trajectory, and then search the optimal solution of the current region. Each update of PSO algorithm, there are two important values need to be retained, one of which is a group of individual particles in the individual optimal solution, denoted as pbest; there is a solution for the global optimal solution, representing the whole search space in the particle swarm optimization solution, denoted as gbest. The velocity and position of the particles are updated according to these two values, and each iteration is updated according to formulas (6) and (7).

$$v_i^{t+1} = w \times v_i^t + c_1 \times r_1 \times (pbest^t - x_i^t) + c_2 \times r_2 \times (gbest^t - x_i^t) \tag{6}$$

$$x_i^{t+1} = x^t + v_i^{t+1} \tag{7}$$

Among them: $i = \{1,2,\ldots SN\}$, other parameters are shown in the following Table 1:

Table 1. Parameters of PSO

Parameters	Explanations of parameters
SN	The size of the particle swarm
w	Tnertia factor, the purpose of which is to balance local optimal solution and global optimal solution
r_1	A random number to ensure the diversity of the whole particle swarm between [0,1]
r_2	A random number between [0,1]
c_1	A learning factor which on behalf of the thinking abilities of particle swarm
c_2	A learning factor which on behalf of the social-behavior ability of particle swarm
V_{max}	The limited speed of particle swarm. The speed would be V_{max} if the speed would be faster than V_{max}

The particle swarm optimization algorithm is applied to some specific application parameters optimization or weight coefficient optimization problems, generally, it can obtain the global optimal solution so that can improve the final experimental results.

2.4 SMOTE and AdaBoost Base on PSO Optimization

Using Borderline algorithm to determine the boundary samples, and then the next step is to carry on under-sampling and over-sampling operation in majority boundary samples and minority boundary samples separately. The main operating procedure is as follows:

First of all, for the minority samples, there are two types of data need to be over sampling, one is the border minority samples which remember as SBmin; the other is the security minority samples which we recorded as SSmin. According to the importance of the sample information, SMOTE were used to over sampling for the data samples in SBmin and SSmin. Using different over sampling rate OsRate according to the importance of the boundary samples and the level of importance of the safety samples.

In the end, the data samples of the majority boundary sample data subset SBmaj are removed from the boundary sample space by some specific under sampling method. By using the specific majority boundary samples, some majority samples are removed to add the range of decisions of minority samples and improve the definition of boundary between the minority samples and the majority samples.

We can get a new data set with clear boundaries between two class and make our classifier get a good result through the above steps.

Resampling method is mainly to change the number of samples of the two types of samples so that the two types of samples tend to balance and can be apply to the traditional classifier for training. In the classification of imbalanced samples, the sample characteristics and the spatial distribution of the samples determine the distribution of the data. In the training process of classification algorithm, if we can choose the most distinguished features, then we will improve the final classification effect, so how to select these important features also play an important role in the imbalance classification problem. Also, the new sample space generated by feature selection can be better represent the characteristics of all data samples while training and reduce the training time of classifier. So in the paper, we use both the over sampling rate and the under sampling rate. Besides, We optimize the features so as to achieve the best data distribution.

By using the particle swarm optimization algorithm, the sampling rate of the boundary samples and the safety samples are optimized to obtain the optimal over sampling rate. At the same time, the features are optimized to select the most representative feature combination to simplify the operation and improve the classification results. AUC/F-Mea is used as the fitness function of the algorithm to improve the final classifier. Particle swarm optimization algorithm in particle representation as shown in Table 2:

Table 2. Particles of PSO

OsRate1	OsRate2N	UsRate	Fec1	FecM

Among them, OsRate represents the over sampling rate, OsRatei is the over sampling rate of the i-th cluster cluster ($i = 1,2,...$ N), N is the number of clusters formed by DBSCAN algorithm; *UsRate* represents the majority class boundary samples of the under sampling rate; *Fec* representative features, *Feci* for the sample of the i-th features ($i = 1,2,...$ M), where the feature vector is binary: 1 on behalf of the location of the feature is to help the classification effect, need to retain while 0 on behalf of the location of the classification results of the upgrade without any help. You can remove this feature to simplify the training process if it is 0.

The pseudo code of the use of PSO algorithm to optimize the over sampling rate in the classification of imbalanced data is shown in Table 3.

Particle swarm optimization algorithm is mainly to optimize the continuous values. Here the feature vectors are discrete. In order to make the PSO can also handle the feature vector, we use a sigmoid function to generate continuous value speed V according to the following formulas (8) and (9).

$$v_i^t = sigmoid\ v_i^t = \frac{1}{1+e^{-v_i^t}} \tag{8}$$

$$x_i^{t+1} = \begin{cases} 1, & r_i < v_i^t \\ 0, & other \end{cases} \tag{9}$$

Table 3. Pseudo code of DBPS

Input: *Dataset*, *MCN*, *SN*, Fitness Function *M*, Sampling step length *step*.
Output: Result of *AUC* or *G-mean*

Divide the Dataset into *Training Set* and *Testing Set*;
Initialize xi(i = 1,2,…,SN);
Set the global optimal solution *gbest* = 0;
for *i* = 1 to *MCN*:
    for *j* = 1 to *SN*:

        Get the solution of x_j

        Count the *pbest* of each particle according to the speed calculation formula (6) and location updating formula (7) of PSO algorithm and update the global optimal solution *gbest* as the optimal *pbest* at the same time.

Use the over sampling rate and under sampling rate geted above to do the mixted sampling. Then train the classifier and test the classifier to get the final *AUC*.

Through the PSO algorithm to determine the sample rate of each cluster cluster and the sampling rate of the safety sample and the feature of the classification, we can extract the most representative data characteristics and data samples, and then through the optimization of the sampling rate of the sample to a small number of samples.

3 Experiments

3.1 Evaluation Standard

It will cause the following problems when using the traditional evaluation criteria considering the special nature of the classification problem: traditional classifier would classify all the minority samples for the majority samples for the pursuit of global classification accuracy. Thus, it would get a high global accuracy but a low recognition rate of minority samples. In this case, the traditional single evaluation system will no longer be applied to the imbalanced classification of the evaluation system. Therefore, we need some special and complex consideration of various indicators to adapt to the special situation of imbalanced sample classification. These standards mainly have two kinds: one is called "the atomic standard", another is called "the compound standard", which is a kind of complex and can be very good to adapt to the imbalanced sample classification problem evaluation system. In addition, the receiver operating characteristic curve (ROC) has been widely used in the evaluation of imbalanced sample classification [23].

The confusion matrix for the two classification problem involved in the classification of imbalanced samples is shown in Table 4. Through the various indicators of statistical confusion matrix and the composite index of these indicators, we can better classify the accuracy of each category respectively. Consider the classification of different categories, so as to evaluate the classification algorithm of imbalanced samples, not blindly pursue the highest accuracy rate, but also consider the classification accuracy of the minority and the majority class.

Table 4. Confusion matrix

	Classified as positive	Classified as negative
Positive	Correct positive TP	Wrong negative FN
Negative	Wrong positive FP	Correct negative TN

Formula (10) to the formula (13) lists some of the commonly used atomic evaluation criteria for the classification of imbalanced samples based on the confusion matrix.

$$Accuracy = 1 - ErrorRate = \frac{TP + TN}{Pc + Nc} \tag{10}$$

$$Precision = \frac{TP}{TP + FP} \tag{11}$$

$$Recall = \frac{TP}{TP + FN} \tag{12}$$

$$F - Measure = \frac{(1 + \beta)^2 \cdot Precision \cdot Recall}{\beta^2 \cdot Recall + Precision} \tag{13}$$

F-Measure is most often applied to the evaluation of imbalanced sample classification, as shown in the formula (13). The F-measure by recall, precision, and to obtain the composite balance factor, when the Recall and Precision are achieved a higher numerical, F-measure will achieve ideal results [24]. Regulation of recall and precision balance factors in the (typically β is 1).

ROC curve (Operating Characteristics Curve Receiver) is proposed by Swets in 1988 [25]. FPRate to ROC for the X axis, TPRate for the Y axis to build the space. By setting a threshold value, get a pseudo positive rate and the ROC curve is formed through the true positive rate values and connect these scattered points.

ROC curve is not able to directly quantify the classification of imbalanced samples, so AUC (Area curve AUC) is proposed in order to get a quantitative evaluation index. The classification effect of classifier algorithm can be evaluated with the area under the right bottom of ROC (that is, AUC). the bigger the AUC is, the better classification effect does the classifier has.

3.2   Data Sets in Experiments

In this paper, we use the experimental data set of eight data sets which are obtained from the UCI database. Table 5 describes the specific attributes of all the data sets used in the experiment. The first column is the data set number, the second column is the name of the dataset, and #Attr is the number of attributes that are included in the data set. %Min represents the proportion of minority class samples. Data set Yeast4 and Yeast5 is about the classification of the yeast protein, Abalone9 is about the classification of abalone's age, Ecoli1 is about the classification of proteins of ecoli cells, Haberman is about the classification of the survival of patients treated with breast cancer surgery, Pima is about the classification of wheter people have the disease.

Table 5. Experimental data set description

No.	Name	#Attr.	% Min.
1	Yeast5	8	2.96
2	Abalone9	8	5.75
3	Glass1	9	8.85
4	Page-blocks0	10	10.23
5	Yeast4	8	10.98
6	Ecoli1	7	22.92
7	Vehicle3	18	25.06
8	Haberman	3	26.47
9	Glass0	9	32.71
10	Pima	8	34.90

3.3   Validate the Verification of SMOTE Base on PSO

Then by comparing the three groups of experiments to verify the effectiveness of the PSO algorithm so as to verify the advancement of the my invention. By comparing the following three experiments to verify the effectiveness of the PSO optimization algorithm:

(A) custom over sampling rate
(B) defined over sampling rate after the specified standard
(C) PSO optimized over sampling rate

Table 6. %ACC result

	%Acc	
A	B	C
97.78	97.93	97.98
92.84	93.84	93.89
90.39	92.51	93.24
95.98	96.69	97.56
93.87	94.47	94.85
89.61	93.69	93.85
78.24	85.47	86.97
73.31	79.73	80.12
82.82	83.41	84.41
74.09	80.71	81.25

Through the results of Table 6, we can get the conclusion: we can get a better results through classify after the PSO optimization of the re sampling rate in the data sample space. It also verifies the effectiveness of the PSO optimization algorithm on the re sampling rate optimization.

Table 7. The result of DBPS

No.	AUC		F-Mea		%Acc	
	DBS	DBPS	DBS	DBPS	DBS	DBPS
1	0.945	**0.953**	**0.732**	0.730	97.93	**97.98**
2	**0.712**	0.703	0.626	**0.659**	93.84	**93.89**
3	**0.640**	0.518	0.276	**0.366**	92.51	**93.24**
4	**0.936**	0.932	0.858	**0.878**	96.69	**97.56**
5	0.893	**0.915**	**0.753**	0.714	94.47	**94.85**
6	0.881	**0.928**	0.824	**0.873**	93.69	**93.85**
7	0.785	**0.842**	**0.659**	0.648	85.47	**86.97**
8	0.711	**0.750**	**0.533**	0.533	79.73	**80.12**
9	0.808	**0.960**	0.759	**0.809**	83.41	**84.41**
10	0.764	**0.860**	0.689	**0.729**	80.71	**81.25**

The third experiments in this section are designed to verify the effectiveness of the PSO algorithm on the over sampling rate optimization. This part of the experiment is about DBPS algorithm which use PSO after the experimental verification of DBS and through the experiment to verify the superiority of the DBPS algorithm.

Table 7 is the result of DBPS. We can be very clear on the advantages and disadvantages of DBS and DBPS algorithm intuitive comparison through the histogram display.

Through the intuitive analysis above Fig. 1, we can get the conclusion: the characteristics of the AUC, F-Measure and %Acc indicators will have a certain improvement

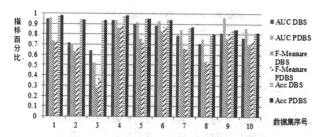

Fig. 1. Comparison between DBS and DBPS

when the PSO algorithm is introduced to optimize the minority over sampling rate. This is mainly because the PSO algorithm can find the global optimal solution while using artificial experience to set the sampling rate is often even not the local optimal solution. So use the PSO algorithm to find the optimal solution of the over sampling rate can get a better classification effect cause the minority samples generated can better represent the original sample space.

In this section, the experiment will be updated from sample weight and integrated classifier Boosting algorithm base classifier optimization decision weighting coefficients of two aspects to improve the effectiveness of the algorithm for the purpose of verify the validity of the algorithm level improvement. Sample weight adjustment scheme mainly reflected in two aspects: the method of weights update of copy samples and the update method of minority samples when training the model after sampling; how to determine the weight coefficients of the base classifiers and how to use PSO to optimize the weight of these base classifiers to do the voting of these classifiers.

In this experiment, AdaBoost.M1 classification algorithm represent Boosting algorithms (AD), SMOTEBoost (expressed as SB) and the algorithm DBPS-Boost proposed in this paper were used. In each experiment results in a higher one with bold font display.

Tables 8 and 9 show that the DBPS-Boost algorithm 6 times achieved the highest AUC value, 8 times achieved the highest F-Mea value, the 9 times get the highest overall classification accuracy. We can get the conclusion through the classification results of these three indicators: after using the new sample weight update method and the PSO algorithm to optimize the weight coefficient of the base classifier, DBPS-Boost can effectively improve the effect of class imbalance classification.

By converting the classification index of all the data sets into the histogram effect, it can better reflect the effectiveness and advanced nature of the DBPS-Boost algorithm in dealing with the problem of classification of imbalanced samples in the class.

Through the summary of the results of the experiments, we can see that using PSO algorithm optimization parameters in DBS over sampling algorithm, also change the weight update method and use PSO to optimize the weight of base classifiers in the algorithm layer can solve these problems well: first of all, minority samples yielded by DBS often have better purposiveness and authenticity and would make the boundary more clear through over sampling the boundary samples. Secondly, the PSO algorithm can get an effective optimal solution to the classification effect. Finally, we use the sample weight adjustment strategy to help ensemble classification learning method to

Table 8. AUC and F-Mea indicator

	AUC			F-Mea		
No	AD	SB	DBPSB	AD	SB	DBPSB
1	0.969	0.970	**0.972**	0.714	0.686	**0.972**
2	**0.807**	0.731	0.785	0.383	0.375	0.785
3	0.796	0.801	**0.823**	0.357	0.286	**0.823**
4	0.989	**0.993**	0.993	0.877	0.823	0.993
5	**0.961**	0.941	0.942	0.755	0.742	0.942
6	0.954	0.953	**0.959**	0.768	0.792	**0.959**
7	0.833	0.838	**0.843**	0.538	0.594	**0.843**
8	0.639	0.650	**0.669**	0.403	**0.479**	**0.669**
9	**0.906**	0.895	0.895	**0.786**	0.722	0.895
10	0.778	0.773	**0.802**	0.594	0.657	**0.802**

Table 9. Chart 10: %*Acc* indicator

No	AD	SB	DBPSB
1	**98.38**	97.78	98.35
2	93.84	90.84	**94.67**
3	90.63	84.39	**95.70**
4	97.48	95.98	**97.59**
5	94.68	93.87	**96.31**
6	89.58	89.61	**94.34**
7	79.67	76.24	**88.22**
8	71.90	68.30	**81.49**
9	85.51	80.82	**88.15**
10	71.74	74.09	**82.64**

put more energy into the original minority samples, and pay more attention to those minority samples that were miss classified but have high credibility. Using PSO to optimize the weight of base classifiers can fully tap the connection between the base classifiers and then find the optimal solution combination of weight of the base classifiers. DBPSB algorithm can effectively improve the classification ability by solving the three main problems when dealing with the imbalance data.

4   Conclusion

Traditional classification algorithm applied to solve the problem of imbalanced data classification often would not have a good result. Most of the researches on the classification of imbalanced samples are mainly based on the data dealing and algorithm improving. Data dealing is mainly through the resampling so that the data set tends to be balanced, the algorithm improving is mainly in the integration of the improved classification algorithm. According to the characteristics of imbalanced samples, this paper improves the data dealing method and classified algorithm, and puts forward a solution to a solution:

(1) In the step of data dealing, taking into account the boundary information of the minority samples, different sampling strategies are adopted for the boundary samples and the security samples. Through the improvement of the above steps, a new algorithm DBS (SMOTE-DBSCAN) is proposed to solve the problem of imbalance-data classification.

(2) In the step of classified algorithm, mainly in the following aspects to improve: first of all in the initial assignment and iterative updates the sample weight was improved based on the original minority samples and modeled on the minority class samples according to their representative of the original sample space and give them different weights respectively.

Finally, the experimental data sets were used to verify the effectiveness of the experimental data sets, using AUC, F-measure and %ACC as the evaluation criteria of the experimental results. From the experimental results, we can see the conclusion: the DBPS algorithm proposed in this paper can get better results when dealing with the problem of sample classification with class imbalance problem.

References

1. Vandenberghe, R., Nelissen, N., Salmon, E., et al.: Binary classification of F-flutemetamol PET using machine learning: comparison with visual reads and structural MRI. NeuroImage **64**, 517–525 (2013)
2. Yun, Z., Nan, Ma., Da, R., et al.: An effective over-sampling for imbalanced data sets classification. Chin. J. Electr. **20**(3), 489–494 (2011)
3. Lazarevic, A., Ertoz, L., Ozgur, A., et al.: Evaluation of outlier detection schemes for detecting network intrusions. In: Proceedings of Third SIAM International Conference on Data Mining, pp. 97–104 (2003)
4. Liu, Y., Chen, Y.: Face recognition using total margin-based adaptive fuzzy support vector machines. Neural Netw. **18**, 178–192 (2007)
5. Kubat, M., Holte, R.C., Matwin, S.: Machine learning for the detection of oil spills in satellite radar images. Mach. Learn. **30**(2), 195–215 (1998)
6. Yin, L., Leong, T.: A model driven approach to imbalanced data sampling in medical decision making. Study Health Technol. Inf. **160**, 856–860 (2010)
7. Huang, Y., Hung, C., Jiau, C.: Evaluation of neural networks and data mining methods on a credit assessment task for class imbalance problem. Nonlinear Anal.: Real World Appl. **7**(4), 720–747 (2006)
8. Vaishali, G.: An overview of classification algorithms for imbalanced data. Emer. Technol. Adv. Eng. **2**(4), 42–47 (2012)
9. Jia, P., Zhang, C., He, Z.: A new sampling approach for classification of imbalanced data sets with high density. In: Proceedings of International Conference on Big Data and Smart Computing, pp. 217–222. IEEE Computer Society (2014)
10. He, H., Garcia, E.: Learning from imbalanced data. Knowl. Data Eng. **21**(9), 1263–1284 (2009)

11. Jo, T., Japkowicz, N.: Class imbalances versus small disjuncts. ACM SIGKDD Explor. Newsl. **6**(1), 40–49 (2014)
12. Chawla, N.V.: Data mining for imbalanced datasets: an overview. Data Min. Knowl. Discov. Handb. 853–867 (2005)
13. Batista, G., Prati, R., Monard, M.: A study of the behavior of several methods for balancing machine learning training data. ACM SIGKDD Explor. Newsl. **6**(1), 20–29 (2004)
14. Kubat, M., Matwin, S.: Addressing the curse of imbalanced training sets: one-sided selection. In: Proceedings of 14th International Conference on Machine Learning, pp. 179–186 (1997)
15. Chawla, N.V., Bowyer, K.W., Hall, L.O., et al.: SMOTE: synthetic minority over-sampling technique. J. Artif. Intell. Res. **16**, 321–357 (2002)
16. Wang, J., Yun, B., Huang, P., Liu, Y.-A.: Applying threshold SMOTE algoritwith attribute bagging to imbalanced datasets. In: Lingras, P., Wolski, M., Cornelis, C., Mitra, S., Wasilewski, P. (eds.) RSKT 2013. LNCS, vol. 8171, pp. 221–228. Springer, Heidelberg (2013)
17. Han, H., Wang, W.-Y., Mao, B.-H.: Borderline-SMOTE: a new over-sampling method in imbalanced data sets learning. In: Huang, D.-S., Zhang, X.-P., Huang, G.-B. (eds.) ICIC 2005. LNCS, vol. 3644, pp. 878–887. Springer, Heidelberg (2005)
18. Maloof, M.: Learning when data sets are imbalanced and when costs are unequal and unknown. In: Proceedings of Conference of Machine Learning, Workshop Learning from Imbalanced Data Sets, pp. 578–597 (2003)
19. Kukar, M., Kononenko, I.: Cost-sensitive learning with neural networks. In: Proceedings of European Conference Artificial Intelligence, pp. 445–449 (1998)
20. Shaoning, P., Lei, Z., Gang, C., Abdolhossein, S., Tao, B., Daisuke, I.: Dynamic class imbalance learning for incremental LPSVM. Neural Netw. **44**, 87–100 (2013)
21. Polikar, R.: Ensemble based systems in decision making. IEEE Circuits Syst. Mag. **6**(3), 21–44 (2006)
22. Galar, M., Fernándezb, A., Barrenechea, E., Bustince, H., Herreraa, F.: A review on ensembles for the class imbalance problem bagging, boosting, and hybrid-based approaches. Syst. Man Cybern. Part C: Appl. Rev. **42**(4), 463–484 (2012)
23. Fawcett, T.: An introduction to ROC analysis. Pattern Recogn. Lett. **27**(8), 861–874 (2006)
24. He, H., Garcia, E.: Learning from imbalanced data. IEEE Trans. Knowl. Data Eng. **21**(9), 1263–1284 (2009)
25. Witten, I., Frank, E.: Data Mining: Practical Machine Learning Tools and Techniques, 3rd edn. Morgan Kaufmann Publishers, USA (2011)

Daily ETC Traffic Flow Time Series Prediction Based on *k*-NN and BP Neural Network

Yanjing Chen[1], Yawei Zhao[1(✉)], and Peng Yan[2]

[1] School of Engineering Science,
University of Chinese Academy of Sciences, Beijing, China
cheneomjing13@mails.ucas.ac.cn,
zhaoyw@ucas.ac.cn
[2] Beijing Sutong Technology Co., Ltd., Beijing, China
yanpeng@ktetc.com

Abstract. Daily Electronic Toll Collection (ETC) traffic flow prediction is one of the fundamental processes in ETC management. The precise prediction of traffic flow provides instructions for transportation hub management solution planning and ETC lane construction. At present, some of studies are proposed in forecasting traffic flow. However, most studies of model presentation are in the form of mathematical expressions, and it is difficult to describe the trend accurately. Therefore, an ETC traffic flow prediction model based on *k* nearest neighbor searching (*k*-NN) and Back Propagation (BP) neural network is proposed, which takes the effect of external factors like holiday, the free of highway and weather etc. into consideration. The traffic flow data of highway ETC lane somewhere is used for prediction. The prediction results indicate that the total average absolute relative error is 5.01 %. The accuracy suggests its advantage in traffic flow prediction and on site application.

Keywords: ETC · Traffic flow prediction · Time series · *k*-NN · BP neural network

1 Introduction

With the rapid development of highway, ETC system, as one of the important subsystems of intelligent transportation system, through ETC technology applied in highway toll modular, effectively improves the traffic capacity of highway, alleviates queuing congestion of artificial charge phenomenon, and what's more, reduces the energy consumption and environmental pollution problems too much. In order to further improve the quality of ETC service and provide instructions for transportation hub management solution planning and ETC lane construction, the research of ETC traffic flow prediction model has important theoretical significance and practical value.

About traffic flow prediction, there have been many researches at home and abroad, and among them, time series model is one of the most mature prediction methods. The most important characteristics of time series model is that it acknowledges the dependency and the correlation between observed values, and it is a kind of dynamic model, so it can be applied to the dynamic prediction [1]. The traditional time series

© Springer Science+Business Media Singapore 2016
W. Che et al. (Eds.): ICYCSEE 2016, Part II, CCIS 624, pp. 135–146, 2016.
DOI: 10.1007/978-981-10-2098-8_17

analysis method is mainly statistical method, like Auto Regression Moving Average (ARMA) and Auto Regressive Integrated Moving Average Model (ARIMA). ARMA model is mainly suitable for linear stationary time series and ARIMA is for difference stationary time series. These are linear approaches. In 1984, Okutani and Stephanedes apply ARIMA to Urban Traffic Control System (UTCS). A lot of improvement models based on the traditional statistics time series prediction models have been proposed. In 2003, Williams and Hoel [2] proposes a seasonal ARIMA process based on the analysis of the shortage of the past traffic flow prediction models, which is performed on the real-world data to verify its effectiveness. Ni et al. [3] firstly decomposes the original traffic flow data into series of time sequence signals that have different characters and then make use of ARIMA to predict. According to the characteristics of urban road short-term traffic flow, Qiu and Yang [4] propose a double seasonal autoregressive integrated moving average model, which can meet the requirements of predicting daily pattern and weekly pattern traffic flow in urban road based on ARIMA and Seasonal Autoregressive integrated moving average model (SARIMA). These time series models based on traditional time series model are mainly in the form of mathematical expression, but because of the complexity and randomness of time series data in the real world, it is difficult to describe the time series accurately by linear model.

The existing research results provide guidance for traffic flow prediction, but it lacks of focus on the research of the influence factors of traffic flow. And as the development and application of data mining and machine learning algorithm, that the k-NN [5], neural network method [6] et al. are applied to traffic flow forecast has attracted more and more attention. Each model has its disadvantages and advantages. Therefore, the combined prediction models become popular. Besides, about the influence factors of traffic flow daily, Zhang et al. [7] find that rainfall will cause certain influence to traffic flow. While the traffic flow prediction model is set up, Cools et al. [8] find the traffic flow is lower during the holiday than the daily traffic volume. The weekdays and the non-workdays of traffic flow are with different characteristics [4]. Meanwhile, predictions based on multidimensional time series prediction model have successfully applied to the prediction of wind power [9], the air quality [10], and weather [11].

This paper proposes a daily ETC traffic flow time series prediction model, which combines k-NN and BP neural network. In order to measure the similarity more accurately and take the external factors of traffic flow into consideration, we use multidimensional time series for k-NN prediction, the dimensions including weather, holiday data etc. The real-world data of ETC lane from February 2011 to August 2015 is used to carry out the experiments to evaluate the feasibility and applicability of the model.

The rest of the paper is organized as follows. Section 2 presents the proposed prediction model combined k-NN and BP neural network. Section 3 shows the evaluation and performance of the comparison and experimental results. The conclusion is given in Sect. 4. The last section is acknowledgment.

2 Methodology

2.1 Multidimensional Time Series

Time series X is a set of dimensions x_i $(i = 1, 2, 3, \ldots, m)$ values in a series of continuous time t_j $(j = 1, 2, 3, \ldots, n)$ orderly. Each time interval of X can be uniform or different, but the present study mainly about uniform time interval. This is, $\Delta t = t_{j+1} - t_j (j = 1, 2, 3, \ldots, n - 1)$ is fixed value. To describe it simply, denote t_j as j, which means the jth time. When $m = 1$, X is a single dimension time series. When $m > 1$, it means x_i contains more than one dimension, and X is called multidimensional time series. X can be expressed by a $m * n$ matrix, denoted as

$$X = \begin{pmatrix} x_{11} & \cdots & x_{1n} \\ \vdots & \ddots & \vdots \\ x_{m1} & \cdots & x_{mn} \end{pmatrix} \tag{1}$$

In (1), x_{ij} means the ith dimension observed value at the jth time. And each row of X is a single dimension time series.

Using sliding window approach to transfer X to a new dataset D. Denote the size of the sliding window as $h + p$. Thus, $D = \{X_1, X_2, \ldots, X_L\}$, $(L = n - h - p + 1)$, and X_j is as follows

$$X_j = \begin{pmatrix} x_{1j} & \cdots & x_{1(j+h+p-1)} \\ \vdots & \ddots & \vdots \\ x_{mj} & \cdots & x_{m(j+h+p-1)} \end{pmatrix}, (1 \le j \le n - h - p + 1) \tag{2}$$

The first h columns of X_j are the input for training models, we call it inw_j. And the last p columns of X_j are the output for prediction model, called $outw_j$. Moreover, denote X_{ij} as the ith row of X_j, which means the time series of the ith dimension and inw_{ij} and $outw_{ij}$ respectively are the input and output of X_{ij}. So multidimensional time series prediction is shown as follows.

$$x'_{i(j+h)} = f_i(inw_j), (1 \le i \le m, 1 \le j \le n - h - p + 1) \tag{3}$$

In this paper, the mth dimension is ETC traffic flow, which is the only target dimension we need to predict. In the following subsections, a model combined k-NN and BP neural network is introduced.

2.2 k-NN

k-NN is based on learning by analogy [12], that is, by comparing a given test sample with training samples that are similar to it. It can be used for regression by returning the average value of the k nearest neighbors. And the merit of k-NN is that it has a few parameters to set and it is easy to explain the result. In this paper, k-NN is used to predict traffic flow using the average of the k nearest neighbors.

k-NN is based on the similarity search. Euclidean distance is simple to compute with linear time complexity, and can keep constant distance under the orthogonal transformation [13]. So in order to measure the similarity of instances, Euclidean distance is used in this paper. On the one hand, as time changes, time series data often can appear the phenomenon of reducing and enlarging, so, with the purpose of reducing the effect of this phenomenon in similarity measure, data normalization is needed before measuring. On the other hand, physical dimension is quite different of dimensions, and the dimension with large scope can cover the small ones. Therefore we choose *Min-Max* normalization method, and each dimension of every *inw* in testing dataset and training dataset is mapped to the interval [0, 1]. Set the timestamps for the training dataset range from 1 to *TL*. The *j*th sample in *D* is X_j, $(1 \leq j \leq TL)$ as shown in (2), and $inw_j = (X_{1j},...,X_{mj})^T$. After inw_j is normalized to the interval [0, 1], having

$$inw'_j = \begin{pmatrix} x'_{1j} & \cdots & x'_{1(j+h-1)} \\ \vdots & \ddots & \vdots \\ x'_{mj} & \cdots & x'_{m(j+h-1)} \end{pmatrix}. \tag{4}$$

$$x'_{i(j+q)} = \frac{x_{i(j+q)} - min_{inw_{ij}}}{max_{inw_{ij}} - min_{inw_{ij}}}, (1 \leq i \leq m, 0 \leq q \leq h-1) \tag{5}$$

In (5), $min_{inw_{ij}}$ and $max_{inw_{ij}}$ are the minimum and maximum values of inw_{ij}.

Given a testing input sample inw_T, starting at time point *T* with length *h*, after normalization, having

$$inw'_T = \begin{pmatrix} x'_{1T} & \cdots & x'_{1(T+h-1)} \\ \vdots & \ddots & \vdots \\ x'_{mT} & \cdots & x'_{m(T+h-1)} \end{pmatrix} \tag{6}$$

$$x'_{i(T+q)} = \frac{x_{i(T+q)} - min_{inw_{iT}}}{max_{inw_{iT}} - min_{inw_{iT}}}, (1 \leq i \leq m, 0 \leq q \leq h-1) \tag{7}$$

In (7), $min_{inw_{iT}}$ and $max_{inw_{iT}}$ are the minimum and maximum values of inw_{iT}.

Then we calculate Euclidean distance between inw_T and each sample in the training dataset, denoted as $Dis(j,T)$ using (8).

$$Dis(j, T) = \sum_{q=0}^{h-1} \sqrt{\sum_{i=1}^{m} (x'_{i(j+q)} - x'_{i(T+q)})^2} \tag{8}$$

Due to the influence of each dimension on the target prediction traffic flow dimension different, add a weight for the distance measure. Therefore, (8) is denoted as

$$Dis(j, T) = \sum_{q=0}^{h-1} \sqrt{\sum_{i=1}^{m} w_i (x'_{i(j+q)} - x'_{i(T+q)})^2} \tag{9}$$

In (9), w_i is the weight of the ith dimension, meaning the correlation between the factor dimension and traffic flow dimension.

K samples with the smallest distance to the testing sample are selected and the corresponding value in $outw$ can be used for prediction. It assumes the jth training samples is the selected neighbor for the Tth testing samples. Because the inw_T and inw_j are similar after normalization, so the corresponding $outw$s are similar after normalization too. Hence, in order to get the predicted traffic flow value, traffic flow dimension in $outw$ needs be normalized.

$$x'_{m(j+q)} = \frac{x_{m(j+q)} - min_{inw_{mj}}}{max_{inw_{mj}} - min_{inw_{mj}}}, (h \leq q \leq p+h-1) \tag{10}$$

$$x'_{m(T+q)} = \frac{x_{m(T+q)} - min_{inw_{mT}}}{max_{inw_{mT}} - min_{inw_{mT}}}, (h \leq q \leq p+h-1). \tag{11}$$

Because $outw_j \approx outw_T$, $x'_{m(j+q)} \approx x'_{m(T+q)}$, Consequently

$$\frac{x_{m(j+q)} - min_{inw_{mj}}}{max_{inw_{mj}} - min_{inw_{mj}}} \approx \frac{x_{m(T+q)} - min_{inw_{mT}}}{max_{inw_{mT}} - min_{inw_{mT}}} \tag{12}$$

And then having

$$x_{m(T+q)} = \frac{x_{m(j+q)} - min_{inw_{mj}}}{max_{inw_{mj}} - min_{inw_{mj}}} (max_{inw_{mT}} - min_{inw_{mT}}) + min_{inw_{mT}} \tag{13}$$

(13) is one final predicted value for traffic flow $x_{m(T+q)}$ in this paper after k-NN. When $k \geq 2$, the average of the k predicted value from (13) is the final predict value. Assuming that $inw_{j1}, \ldots, inw_{jk}$ are the k nearest neighbors for inw_T, the final predicted value usually can be computed using the arithmetic mean and the weighted mean. To the arithmetic mean, the predicted traffic flow result is denoted as follows:

$$\hat{x}_{m(T+q)} = \frac{\sum_{c=1}^{k} \hat{x}_{m(T+q)_{jc}}}{k} \tag{14}$$

In (14), $\hat{x}_{m(T+q)_{jc}}$ is the predicted traffic flow value from $X_{m(jc)}$.

To the weighted mean, the predicted traffic flow result is denote as

$$\hat{x}_{m(T+q)} = \sum_{c=1}^{k} w_{jc} \hat{x}_{m(T+q)_{jc}} \tag{15}$$

In (15), w_{jc} is:

$$w_{jc} = \frac{Dis(jc, T)^{-1}}{\sum_{a=1}^{k} Dis(ja, T)^{-1}} \tag{16}$$

2.3   BP Neural Network

In recent years, the neural network model has been used in different fields. Because the traffic system is always dynamic and complex, it can't be describe perfectly with deterministic linear function, but neural network can solve it. In this paper, we use BP neural network model to predict traffic flow and its structure is shown in Fig. 1.

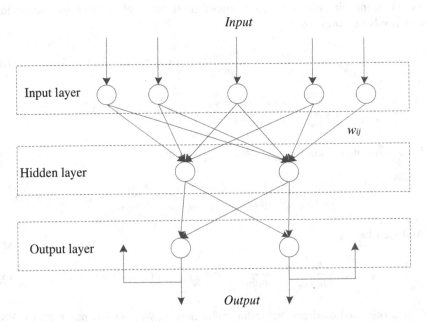

Fig. 1. BP neural network model structure

The BP neural network, as shown in Fig. 1, has three kinds of layers, including the input layer, the hidden layer and the output layer. The number of hidden layer can be greater than one. But with the increase of layer number, it will be more complex to train. The training procedure is:

(1) Original data collecting and preprocessing, such as reconstruction and normalization.
(2) Designing network topology, including the number of hidden layer, the number of the input layer nodes, the number of the hidden layer nodes and the number of the output layer node.
(3) Dividing the data from (1) into training dataset and testing dataset.
(4) Designing learning algorithm and training the BP neural network prediction model.
(5) Combing training datasets with given error to evaluate the BP neural network model performance. When the number of training reaches the given max number of training, go into step (6). If present accumulated errors are greater than given error threshold, return (4); if not, go to (6).
(6) Importing testing dataset to BP neural network model that satisfies training requirement, and we can get the corresponding prediction values.

In this paper, the *m*th dimension is the target prediction traffic flow dimension, so we use the inw_{mj} and $outw_{mj}$ as a pair of input and output. Besides, Levenberg-Marquadt algorithm is employed here.

2.4   Proposed Method

The combined prediction model is using various methods and integrating their features in some way to get the best prediction result finally. The proposed method combines *k*-NN and BP neural network, so that, we can get two prediction results from them. In order to construct the best nonlinear combination model of prediction, reusing BP neural network to train the prediction model with the two predicted results as input. Figure 2 shows the model framework of our proposed method combining *k*-NN and BP neural network.

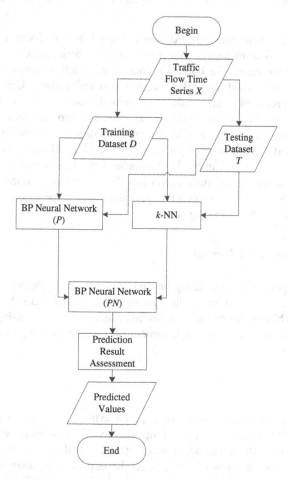

Fig. 2. Model framework of the proposed method

As it is shown in Fig. 2, the steps of our proposed method are as follows:

(1) Using the sliding window approach to transfer the traffic flow multivariate time series X to training dataset D and testing dataset T.
(2) Using k-NN to find the k nearest neighbors of samples in D. Every time a search is achieved, adding the sample to the searching dataset.
(3) Employing BP neural network to train a prediction model based on D and denoting this network model as P.
(4) Combining the prediction results from (2) and (3) as input and the real values as output to train the BP neural network and denoting it as PN.
(5) Taking every sample in testing dataset T to prediction model P and k-NN to get corresponding prediction results.
(6) Applying the results from (5) to prediction model PN as input data and temp predicted result we can get.
(7) Validating and modifying the prediction results from (6) with known rules, like boundary constraint, and the validated values are the final predicted values.
(8) Ending.

The multidimensional time series data is collected from real-world traffic system, which contains physical meanings. According to the experience or expert advice, we can get some known rules, like boundary constraint, which means that the values of time series data value would fall within a certain reasonable range. Using known rules to reprocess the predicted values makes the method more reliable. In this paper, we know traffic flow is equal or greater than 0, so when the predicted result is less than 0, use the average value of traffic flow in the corresponding *inw* as the final result.

In our proposed method, a set of parameters should be initialized, like the length of *inw h*, the length of *outw p*, k in the k-NN approach etc. In this paper, we mainly talk about $p = 1$ and the number of hidden layer is 1. The other values of the parameters are selected combined with the result of experiments and experience, which will be explained in detail in the following section.

3 Experiments and Results

To evaluate the performance of the proposed method, we conducted various experiments to compare its prediction results with the k-NN model and BP neural network. We denote the k-NN model based on multidimensional time series using (14) as knn1 and (15) as knn2, and based on single dimension time series using (14) as knn3 and (15) as knn4.

3.1 Dataset

The comparative experiments are carried out based on the ETC lane daily traffic flow data from February 2011 to August 2015, and the other five dimensions are weather, holiday and so on in the meantime. The detail information of the dataset is shown in Table 1. The ETC traffic flow is shown in Fig. 3. In the experiments, after sliding window approach, the data before 2015 is used for training and the data in 2015 is for testing.

Table 1. Dataset of ETC lane daily traffic flow

Dimension	Data type	Meaning
Traffic_date	Date	Traffic date
Day_of_week	Int	Day of week, like Tuesday denoted as 2
Isholiday	Int	Whether it is holiday. If yes, denote 1, otherwise 0
Isfree	Int	Whether it is highway free. If yes, denote 1; otherwise 0
Weather	Int	Weather conditions, for example, the sunny day is 0, cloudy day is 2, foggy day is 3 etc.
Windforce	Int	The force of wind
Count	Int	Daily traffic flow

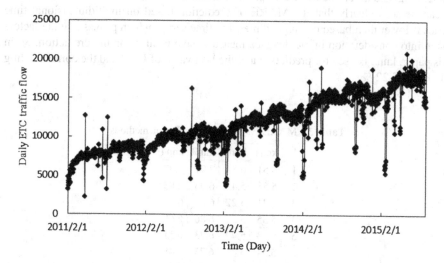

Fig. 3. Time series diagram of highway ETC lane traffic flow

3.2 Error Measure

The prediction performance is quantified by Relative Error (RE) and Mean Absolute Relative Error (MARE). The RE and MARE are defined as follows:

$$RE(t) = \frac{x_t - \widehat{x}_t}{x_t} \tag{17}$$

$$MARE = \frac{1}{N}\sum\nolimits_{t=1}^{N} |RE(t)| * 100\% \tag{18}$$

Where N is the number of instances to be predicted, $\widehat{x}_t$ is the predicted value, x_t the real value. The smaller MARE is, the more precise it predicts.

3.3   Experimental Results

As mentioned above, a group of parameters need to be determined. The value ranges of parameter for the datasets we use are listed in Table 2.

Table 2. Value ranges for parameter tuning

h	p	k
[3, 15]	[1, 5]	[1, 20]

When set $h = 8$, we compare the prediction results of different k values. The MARE values of different methods are shown in Table 3. Through Comparing the knn1 and knn2, knn3 and knn4, the MARE of prediction using (15) is lower than using (14). And it can be seen clearly that the MARE of prediction based on multidimensional time series is lower than based on single dimension time series, which proves that the factors taken into consideration in the distance measure have positive on the prediction. So in this paper, knn2 is used for prediction and the best value of k is 3 and the corresponding MARE is 5.22 %.

Table 3. MARE (%) values of different methods

k	knn1	knn2	knn3	knn4
1	6.51	6.51	7.97	7.97
2	5.55	5.43	6.96	6.83
3	5.21	5.22	6.86	6.63
4	5.55	5.54	6.86	6.65
5	5.43	5.42	6.73	6.51
6	5.54	5.46	6.73	6.48
7	5.53	5.44	6.63	6.40
8	5.55	5.46	6.45	6.27
9	5.66	5.55	6.30	6.16
10	5.61	5.47	6.28	6.13
11	5.61	5.47	6.25	6.09
12	5.58	5.46	6.19	6.03
13	5.53	5.43	6.23	6.02
14	5.57	5.45	6.37	6.08
15	5.63	5.50	6.40	6.10
16	5.71	5.55	6.48	6.12
17	5.82	5.62	6.53	6.14
18	5.82	5.63	6.55	6.16
19	5.78	5.60	6.60	6.19
20	5.86	5.68	6.67	6.23

To the BP neural network model, the parameters we use are as follows: the number of hidden layer is 1and the number of hidden layer node is 5. Besides, the activation function of the hidden layer is sigmoid function. The values of MARE of the three approaches, including knn2, BP neural network and the proposed method, are shown in Table 4. As can be seen from the table, the proposed method has lower MARE than the other two methods. After comparing the MARE, the proposed method has lower MARE than the other two methods. At the same time, the prediction results of the proposed method are shown in Fig. 4.

Table 4. Performance in terms of MARE (%)

knn2	BP neural network	The proposed method
5.22	5.51	5.01

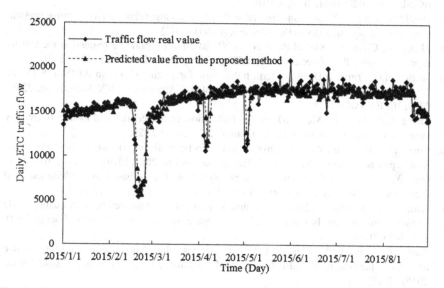

Fig. 4. Comparison of the predicted result from the proposed method and the real ETC traffic flow

4   Conclusion

In this paper, we propose a time series prediction based on *k*-NN and BP neural network to predict ETC traffic flow, which integrate the merit of the two methods. The factors which influence traffic flow are considered, so *k*-NN component is used to make full use of the historical data to predict. Besides, *k*-NN has a few parameters and it easy to explain the predicted results. Meanwhile, the predicted values assessment component is added to reprocess the predicted values to make sure that the predicted result has a more reliable physical meaning. The results show that our proposed method has a better performance and lower prediction error.

Acknowledgments. The author would like to thank the members of the team for providing the helpful discussions and ideas. In addition, we would like to thank every teacher that provides instruction.

References

1. Han, C., Song, S., Wang, C.H.: A real-time short-term traffic flow adaptive forecasting method based on ARIMA model. J. Syst. Simul. **16**(7), 1530–1532 (2004)
2. Williams, B.M., Hoel, L.A.: Modeling and forecasting vehicular traffic flow as a seasonal ARIMA process: theoretical basis and empirical results. J. Transp. Eng. **129**(6), 664–672 (2003)
3. Ni, L., Chen, X., Huang, Q.: ARIMA model for traffic flow prediction based on wavelet analysis. In: 2010 2nd International Conference on Information Science and Engineering (ICISE), pp. 1028–1031. IEEE (2010)
4. Qiu, D.G., Yang, H.Y.: A short-term traffic flow forecast algorithm based on double seasonal time series. J SiChuan Univ. Eng. Sci. Ed. 45(05) (2013)
5. Zhang, T., Chen, X., Xie, M.P., et al.: K-NN based nonparametric regression method for short-term traffic flow forecasting. Syst. Eng. Theory Pract. **30**(2), 376–384 (2010)
6. Huang, H.Q., Tang, T.H.: Short-term traffic flow forecasting based on ARIMA-ANN. In: 2007 IEEE International Conference on Control and Automation, ICCA 2007, pp. 2370–2373. IEEE (2007)
7. Zhang, C.B., Wan, P., Mei, C.H., et al.: Traffic flow characteristics and models of freeway under rain weather. J. Wuhan Univ. Technol. **35**(3), 63–67 (2013)
8. Cools, M., Moons, E., Wets, G.: Investigating effect of holidays on daily traffic counts: time series approach. Transp. Res. Rec. J. Transp. Res. Board **2007**(2019), 22–31 (2019)
9. Cao, X.L., Wang, S.P.: Short time wind power prediction based on multidimensional time-series and BP neural networks. Shanxi Electr. Power **42**(4), 19–23 (2014)
10. Kadiyala, A., Kumar, A.: Multivariate time series models for prediction of air quality inside a public transportation bus using available software. Environ. Prog. Sustain. Energy **33**(2), 337–341 (2014)
11. Bueno, L., Costa, P., Mendes, I., et al.: Evolving ensemble of fuzzy models for multivariate time series prediction. In: IEEE International Conference on Fuzzy Systems (FUZZ-IEEE 2015) (2015)
12. Han, J.W.: Data Mining Concept and Techniques. China Machine Press, Beijing (2006)
13. Ferhatosmanoglu, H., Tuncel, E., Agrawal, D., et al.: Approximate nearest neighbor searching in multimedia databases. In: International Conference on Data Engineering, pp. 503–511. IEEE Computer Society (2001)

Integrity Constraint Validation in *DL-Lite$_R$* Based Ontology Using Rewriting

Xianji Cui[1,2], Dantong Ouyang[2], and Jialiang He[1(✉)]

[1] College of Information and Communication Engineering, Dalian Minzu University, Dalian 116600, People's Republic of China
cuixianji@dlnu.edu.cn, peak_email@sina.com
[2] Key Laboratory of Symbolic Computation and Knowledge Engineering, Ministry of Education, Jilin University, Changchun 130012, People's Republic of China
ouyd@jlu.edu.cn

Abstract. With the rapid development of information technology, semantic web data present features of massiveness and complexity. As the data-centric science, social computing have great influence in collecting and analyzing semantic data. In our contribution, we propose an integrity constraint validation for *DL-Lite$_R$* based ontology in view of data correctness issue in the progress of social computing applications. Firstly, at the basis of translations from integrity constraint axioms into a set of conjunctive queries, integrity constraint validation is converted into the conjunctive query answering over knowledge bases. Moreover, rewriting rules are used for reformulating the integrity constraint axioms using standard axioms. On this account, the integrity constraint validation can be reduced to the query evaluation over the ABox, and use query mechanisms in database management systems to optimize integrity constraint validation. Finally, the experimental result shows that the rewriting-based method greatly improves the efficiency of integrity constraint validation and is more appropriate to scalable data in the semantic web.

Keywords: Semantic web · Social computing · Integrity constraints · Conjunctive query · Rewriting

1 Introduction

Semantic web is an extension of the WWW in which information is expressed in intelligent forms, better enabling computers to understand. Ontology is used as the conceptual schema to represent data in the semantic web, and Web Ontology Language (OWL) [1] is a candidate recommendation ontology language for the W3C. Social Computing is a kind of intelligent computing model that allows users to participate in managing Web data by giving a brief and rich personalized labels in the form of social tagging [2]. It may improve the efficiency of resource retrieval and identification without any professional knowledge.

However, these labeled data cannot be guaranteed to be correct, and it is necessary to validate these labeled data. There are several mature tools For the

© Springer Science+Business Media Singapore 2016
W. Che et al. (Eds.): ICYCSEE 2016, Part II, CCIS 624, pp. 147–156, 2016.
DOI: 10.1007/978-981-10-2098-8_18

general syntax errors and logical conflicts in Semantic Web. But for the most data-centric applications, it also need to consider the user constraints [3,4]. Take WeChat group as an example, we need to limit the number of people, the number of words, etc. These constraints need to be expressed in the form of rules, and automatically verified and repaired also in other social computing applications. Therefore, it is needed to introduce integrity constraints into the ontology. Integrity constraints were originally proposed in databases and artificial intelligence knowledge representation languages to guarantee legal states which are considered acceptable by knowledge bases. They could be added into the ontology to guarantee these legal states of data in OWL ontologies [5].

There is a long research tradition in this work. Several existing proposals model integrity constraints with formulas with epistemic operators [6] and non-monotonic rules [7]. However, these modelings have modified the syntax of standard DL knowledge bases, and no longer appropriate to do the standard reasoning in DL knowledge bases. Based on this point, researchers aim to modify the semantics of axioms in DL knowledge bases to meet the semantics of integrity constraints, and using minimal models to define the satisfaction of integrity constraints [8]. In this paper, we agree with this point, and define the satisfaction of integrity constraints with minimal models. Our closest related work proposed by Tao et al. [9] based on the idea that convert the integrity constraint validation into the SPARQL query answering. However, since SPARQL queries only focus on ABox instances and ignore the standard reasoning, which is the characteristic of standard DL knowledge bases. It may lose results of standard reasoning in DL knowledge bases which is important in the integrity constraint validation. In view of above-mentioned facts, we aim to propose an integrity constraint mechanism by rewriting technology to capture the intuition of integrity constraints and improve the validation of integrity constraints for $DL\text{-}Lite_R$.

2   Integrity Constraint Knowledge Bases

As the logical foundation of OWL, Description Logics (DLs) [10] have provided a sound and complete reasoning algorithm for the standard reasoning in knowledge bases. To extend standard description logic knowledge bases with integrity constraints, an extended knowledge base has been defined as follows.

Definition 1. *Call an extended description logic knowledge base* $<\mathcal{K}, \mathcal{IC}>$ *be an integrity constraint knowledge base (IC-KB), where* $\mathcal{K}$ *is a standard knowledge base that represents the domain of interest, and* $\mathcal{IC}$ *is the set of integrity constraint axioms used for checking the integrity of* $\mathcal{K}$.

In this paper, we concentrate on the integrity constraint validation for $DL\text{-}Lite_R$ knowledge bases. The $DL\text{-}Lite$ family of description logics are a set of tractable DLs which are designed with the specific goal of allowing the reasoning for a large amount of instance data, and $DL\text{-}Lite_R$ is a DL of $DL\text{-}Lite$ family with role inclusions. The syntax and semantics of $DL\text{-}Lite_R$ is detailed in [11]. Then, the satisfaction of integrity constraints in $DL\text{-}Lite_R$ based ontology is defined based on the idea of minimal models: a constraint axiom is satisfied if all minimal models satisfy it. The set of minimal models of $\mathcal{K}$ is denoted with $Mod_M(\mathcal{K}) = \{\mathcal{I} \in Mod(\mathcal{K}) \mid \nexists \mathcal{J}, \mathcal{I} \in Mod(\mathcal{K}), \mathcal{J} \subset \mathcal{I}\}$.

Definition 2. *Let* $<\mathcal{K}, \mathcal{IC}>$ *be an IC-KB, for every IC-axiom* $\alpha \in \mathcal{IC}$, *we say that* $\mathcal{K}$ *satisfies* α, *denoted with* $\mathcal{K} \models_{IC} \alpha$ *if and only if for every minimal model* $\mathcal{I}$ *of* $\mathcal{K}$, $\mathcal{I} \models \alpha$. *Further, we say that* $\mathcal{K}$ *satisfies* $\mathcal{IC}$, *if* $\mathcal{K} \models_{IC} \alpha$ *for every IC-axiom* $\alpha \in \mathcal{IC}$.

3   Integrity Constraint Validation

In this section, we start by translating IC-axioms to conjunctive queries, and then describe how to reformulate corresponding queries with standard axioms, give the integrity checking algorithm and finally prove the correctness.

3.1   IC Validation into Conjunctive Query Answering

To adhere to the closed world semantics in integrity constraint validation, Glimm et al. have added the negation as failure denoted with "**not**" into conjunctive queries [12]. With this notion of query, a conjunctive query over a KB $\mathcal{K}$ is a conjunctive query whose atoms are defined as follows:

$at \leftarrow A(z) \mid R_a(z_1, z_2) \mid \textbf{not } A(z) \mid \textbf{not } R_a(z_1, z_2) \mid z_1 = z_2 \mid \textbf{not } z_1 = z_2$

where A and R_a are respectively an atomic concept and an atomic role of $\mathcal{K}$, and z_i are either constants in $\mathcal{K}$ or variables.

Definition 3. *Let q be a conjunctive query,* $\boldsymbol{a}$ *be a tuple of constraints appearing in* $\mathcal{K}$. *We say* $\boldsymbol{a}^\mathcal{I} \in q^\mathcal{I}$ *if, for each atom* $at_i \in q$, *there exists* $at_i^\mathcal{I} \subseteq \boldsymbol{a}$, *for each model* $\mathcal{I}$ *of* $\mathcal{K}$, *where*

$$at_i^\mathcal{I} = \begin{cases} \{a_1\} & \text{if } at_i = A(z), \exists a_1 \in N_k \text{ s.t. } a_1^\mathcal{I} \in A^\mathcal{I} \\ & \text{if } at_i = \textbf{not } A(z), \exists a_1 \in N_k \text{ s.t. } a_1^\mathcal{I} \notin A^\mathcal{I} \\ \{a_1, a_2\} & \text{if } at_i = R_a(z_1, z_2), \exists a_1, a_2 \in N_k \text{ s.t. } (a_1^\mathcal{I}, a_2^\mathcal{I}) \in R_a^\mathcal{I} \\ & \text{if } at_i = \textbf{not } R_a(z_1, z_2), \exists a_1, a_2 \in N_k \text{ s.t. } (a_1^\mathcal{I}, a_2^\mathcal{I}) \notin R_a^\mathcal{I} \\ \{\varnothing\} & \text{other cases} \end{cases} \quad (1)$$

Definition 4. *Given a query q and a KB* $\mathcal{K}$, *the answer to q over* $\mathcal{K}$ *is the set* $ans(q, \mathcal{K})$ *of tuples* $\boldsymbol{a}$ *of constraints appearing in* $\mathcal{K}$ *such that* $\boldsymbol{a}^\mathcal{I} \in q^\mathcal{I}$ *for every model* $\mathcal{I}$ *of* $\mathcal{K}$.

Note also that the tuple $\boldsymbol{a}$ can be the empty tuple. In this case we say that the answer to q over $<\mathcal{K}$ is null, denoted by $ans(q, \mathcal{K}) = \varnothing$.

Then IC-axioms are translated into the union of conjunctive queries with "**not**" and further converted into conjunctive queries by splitting disjunctions in the body of query into different queries with the same head. It is well-known that all nowadays DL languages are subsets of DL $\mathcal{SROIQ}$ [13]. Thus, integrity constraint axioms (IC-axioms) represented with expressive DL $\mathcal{SROIQ}$ may preferably capture the meaning of integrity constraints. Translation rules are shown in

Table 1. Translation rules from IC-axioms to conjunctive queries with "**not**"

DL	UCQ
$\pi_c(A, x)$	$A(x)$
$\pi_c(\{a\}, x)$	$x = a$
$\pi_c(\neg C, x)$	$\textbf{not } \pi_c(C, x)$
$\pi_c(C_1 \sqcap C_2, x)$	$\pi_c(C_1, x) \wedge \pi_c(C_2, x)$
$\pi_c(\geq nS.C, x)$	$\bigwedge_{i=1}^{n}(\pi_r(S, x, y_i) \wedge \pi_c(C, y_i)) \bigwedge_{i=1}^{n} {}_{j=i+1}^{n}\textbf{not } (y_i = y_j)$
$\pi_r(R, x_1, x_2)$	$R(x_1, x_2)$
$\pi_r(R^-, x_1, x_2)$	$R(x_2, x_1)$
$\pi(C_1 \sqsubseteq C_2)$	$\pi_c(C_1, x) \wedge \textbf{not } \pi_c(C_2, x)$
$\pi(R_1 \circ ... \circ R_n \sqsubseteq R)$	$\pi_r(R_1, x_1, x_2) \wedge ... \wedge \pi_r(R_n, x_{n-1}, x_n) \wedge \textbf{not } \pi_r(R, x_1, x_n)$
$\pi(\mathrm{Trans}(R))$	$\pi_r(R, x_1, x_2) \wedge \pi_r(R, x_2, x_3) \wedge \textbf{not } \pi_r(R, x_1, x_3)$
$\pi(\mathrm{Ref}(R))$	$\textbf{not } (\pi_r(R, x, x))$
$\pi(\mathrm{Asy}(R))$	$\pi_r(R, x_1, x_2) \wedge \pi_r(R, x_2, x_1)$
$\pi(\mathrm{Dis}(R, S))$	$\pi_r(R, x_1, x_2) \wedge \pi_r(S, x_1, x_2)$

where $A, C_{(i)}$ are concepts, $R_{(i)}, S$ roles, a, b individuals, $x_{(i)}, y_{(i)}$ variables

Table 1, where π_c, π_r, π for translate concepts, roles and axioms, respectively. It follows the idea that translate an IC-axiom into a union conjunctive query such that when the query over $\mathcal{K}$ is null the IC-axiom is satisfied, otherwise violated.

According to the translation of IC-axioms into conjunctive queries, we present the theorem which shows that the conversion of integrity constraint validation into conjunctive query answering under closed world assumption is sound and complete.

Theorem 1. *Given an IC-KB* $<\mathcal{K}, \mathcal{IC}>$*, for each* $\alpha \in \mathcal{IC}$*, we say that* $\mathcal{K} \models_{\mathrm{IC}} \alpha$ *if and only if* $ans(\pi(\alpha), \mathcal{K}) = \varnothing$*.*

Proof. We only show the proof for the axiom $C_1 \sqsubseteq C_2$, proofs for other axioms are similar.

$(\Longrightarrow)$: $\mathcal{K} \models_{\mathrm{IC}} C_1 \sqsubseteq C_2$
$\Longrightarrow$ for every $\mathcal{I} \in Mod_M(\mathcal{K}), \mathcal{I} \models C_1 \sqsubseteq C_2$
$\Longrightarrow$ for every $\mathcal{I} \in Mod_M(\mathcal{K}), C_1^{\mathcal{I}} \subseteq C_2^{\mathcal{I}}$
$\Longrightarrow$ for every $\mathcal{I} \in Mod_M(\mathcal{K})$ and $\forall a \in N_k$, if $a^{\mathcal{I}} \in C_1^{\mathcal{I}}$ then $a^{\mathcal{I}} \in C_2^{\mathcal{I}}$
Assume $ans(\pi(C_1 \sqsubseteq C_2), \mathcal{K}) \neq \varnothing$
$\Longrightarrow ans(C_1(x) \wedge \textbf{not } C_2(x), \mathcal{K}) \neq \varnothing$
$\Longrightarrow$ for every $\mathcal{J} \in Mod(\mathcal{K}), \exists a \in N_k$ s.t. $ans(C_1(x) \wedge \textbf{not } C_2(x), \mathcal{K}) = a$
$\Longrightarrow$ for every $\mathcal{J} \in Mod(\mathcal{K}), a^{\mathcal{J}} \in (C_1(x) \wedge \textbf{not } C_2(x))^{\mathcal{J}}$
$\Longrightarrow$ for every $\mathcal{J} \in Mod(\mathcal{K}), \exists \sigma : x \to a, s.t. a^{\mathcal{J}} \in (C_1(a) \wedge \textbf{not } C_2(a))^{\mathcal{J}}$
$\Longrightarrow$ for every $\mathcal{J} \in Mod(\mathcal{K}), a^{\mathcal{J}} \in C_1^{\mathcal{J}}, a^{\mathcal{J}} \notin C_2^{\mathcal{J}}$
$\Longrightarrow$ for every $\mathcal{I} \in Mod_M(\mathcal{K}), a^{\mathcal{I}} \in C_1^{\mathcal{I}}, a^{\mathcal{I}} \notin C_2^{\mathcal{I}}$, since $Mod_M(\mathcal{K}) \subseteq Mod(\mathcal{K})$
$\Longrightarrow$ for every $\mathcal{I} \in Mod_M(\mathcal{K}), a^{\mathcal{I}} \in C_1^{\mathcal{I}}$ and $a^{\mathcal{I}} \notin C_2^{\mathcal{I}}$, which yields a contradiction.
$(\Longleftarrow)$: Assume to the contrary, $\mathcal{K} \not\models_{\mathrm{IC}} C_1 \sqsubseteq C_2$
$\Longrightarrow \exists \mathcal{I} \in Mod_M(\mathcal{K})$ such that $\mathcal{I} \not\models C_1 \sqsubseteq C_2$
$\Longrightarrow \exists \mathcal{I} \in Mod_M(\mathcal{K}) \exists a \in N_k$ such that if $a^{\mathcal{I}} \in C_1^{\mathcal{I}}$, then $a^{\mathcal{I}} \notin C_2^{\mathcal{I}}$

But according to the condition, $ans(\pi(C_1 \sqsubseteq C_2), \mathcal{K}) = \varnothing$
$\Longrightarrow ans(C_1(x) \wedge \textbf{not } C_2(x), \mathcal{K}) = \varnothing$
$\Longrightarrow \forall \mathcal{J} \in Mod(\mathcal{K}) \; \nexists \sigma : x \rightarrow a, \; s.t. \; a^{\mathcal{J}} \in (C_1(x) \wedge \textbf{not } C_2(x))^{\mathcal{J}}$
$\Longrightarrow \forall \mathcal{J} \in Mod(\mathcal{K}), \forall a \in N_k, \text{ if } a^{\mathcal{J}} \in C_1^{\mathcal{J}} \text{ then } a^{\mathcal{J}} \in C_2^{\mathcal{J}}$
$\Longrightarrow \forall \mathcal{I} \in Mod_M(\mathcal{K}), \forall a \in N_k, \text{ if } a^{\mathcal{I}} \in C_1^{\mathcal{I}} \text{ then } a^{\mathcal{I}} \in C_2^{\mathcal{I}}$
(since $Mod_M(\mathcal{K}) \subseteq Mod(\mathcal{K})$), which yields a contradiction.

By doing this, the integrity constraint validation can be converted into the query answering over knowledge bases. Several efforts have been spent towards query answering over the expressive DL knowledge bases [11,12]. However, the computational complexity of the query answering based on these approaches is rather high. In the following section, we will propose the rewriting algorithm to optimize the integrity constraint validation.

3.2   IC Validation Based on Rewriting

Rewriting is an important technique in automated reasoning [14]. It captures the property that reduce query answering into evaluating query over the ABox $\mathcal{A}$ considered as a set of database instances. Given standard axioms $\mathcal{T}$ expressed with *DL-Lite$_R$*, the rewriting algorithm compile one query q into a finite set of queries with the TBox to simulate the evaluation of the query over the whole knowledge base $\mathcal{K}$ by evaluating these rewritten queries over the initial ABox.

The query rewriting process is shown in the following. A standard axiom β in $\mathcal{T}$ is applicable to an atom g of query q if g and right-hand side of standard axiom β has the same predicate (concept or role) name. The atom obtained from the atom g is indicated with $gr(g, \beta)$ by applying the applicable axiom β. Rewriting rules are as follows:

1. For the concept inclusion axiom
   - if the predicate in left-hand side of β is A, then $gr(g, \beta) = A(x)$
   - if $\exists P$, then $gr(g, \beta) = P(x, y)$
   - if $\exists P^-$, then $gr(g, \beta) = P(y, x)$.
2. For the role inclusion axiom
   - if $g = P(x_1, x_2)$ and either β is $P_1 \sqsubseteq P_2$ or $P_1^- \sqsubseteq P_2^-$, then $gr(g, \beta) = P_1(x_1, x_2)$
   - if $g = P(x_1, x_2)$ and either β is $P_1 \sqsubseteq P_2^-$ or $P_1^- \sqsubseteq P_2$, then $gr(g, \beta) = P_1(x_2, x_1)$.

For each conjunctive query q and standard axiom β in $\mathcal{T}$, if there exists an atom in q s.t. β is applicable to g then substitute g with $gr(g, \beta)$. Further, it is necessary to eliminate redundant atoms. That is, for each atom g_1, g_2 in q, if g_1 and g_2 unify, do the most general unifier to q between g_1 and g_2. Iteratively apply this rule until there is no axioms in $\mathcal{T}$ appropriate to g of query q.

We then normalize the *DL-Lite$_R$* knowledge base according to the following transforming rules. Since all axioms in $\mathcal{T}$ are of the principle conjunction normal form, for each positive inclusion with conjunctive concepts is rewritten by iterative application of the rule: if $B \sqsubseteq C_1 \sqcap C_2$ occurs in $\mathcal{T}$, then replace it with two assertions $B \sqsubseteq C_1$ and $B \sqsubseteq C_2$. The standard axioms in $\mathcal{T}$ are expanded by computing all negative inclusions by the following inference rule:

1. if $B_1 \sqsubseteq B_2$ occurs in $\mathcal{T}$ and either $B_2 \sqsubseteq \neg B_3$ or $B_3 \sqsubseteq \neg B_2$ occurs in $\mathcal{T}$, then add $B_1 \sqsubseteq \neg B_3$ to $\mathcal{T}$.
2. if $R_1 \sqsubseteq R_2$ occurs in $\mathcal{T}$ and either $R_2 \sqsubseteq \neg R_3$ or $R_3 \sqsubseteq \neg R_2$ occurs in $\mathcal{T}$, then add $R_1 \sqsubseteq \neg R_3$ to $\mathcal{T}$.
3. if $R_1 \sqsubseteq R_2$ occurs in $\mathcal{T}$ and either $\exists R_2 \sqsubseteq \neg B$ or $B \sqsubseteq \neg \exists R_2$ occurs in $\mathcal{T}$, then add $R_1 \sqsubseteq \neg B$ to $\mathcal{T}$.
4. if $R_1 \sqsubseteq R_2$ occurs in $\mathcal{T}$ and either $\exists R_2^4 \sqsubseteq \neg B$ or $B \sqsubseteq \neg \exists R_2^-$ occurs in $\mathcal{T}$, then add $R_1^- \sqsubseteq \neg B$ to $\mathcal{T}$.

Then, the whole algorithm IC_SATISFY is described as follows to check the satisfaction of integrity constraints for $DL\text{-}Lite_R$ knowledge base.

Algorithm 1. IC_SATISFY($\mathcal{T}$, $\mathcal{A}$, $\mathcal{IC}$)

Input:
   A set of standard axioms $\mathcal{T}$, A set of assertions $\mathcal{A}$, A set of IC-axioms $\mathcal{IC}$
Output:
   $satis$ if each IC-axiom $\alpha \in \mathcal{IC}$ is satisfied return true, otherwise false
1: Initialize $satis \leftarrow$ true
2: Normalize $\mathcal{T}$
3: If $\mathcal{IC}$ is null, return $satis$, for each IC-axiom $\alpha \in \mathcal{IC}$, $\mathcal{IC} \leftarrow \mathcal{IC} \backslash \{\alpha\}$
4: Translate α into conjunctive queries CQ
5: If CQ is null, go to step3. for each query $cq \in CQ$, $CQ \leftarrow CQ \backslash cq$
6: If $ans(cq, \mathcal{A})$ is null, go to step5. Rewrite each query cq with axioms in $\mathcal{T}$ to get new queries Qr, $Qr \leftarrow Qr \backslash \{cq\}$, if Qr is null return false
7: If Qr is null go to step 5, for each query $qr \in Qr$, $Qr \leftarrow Qr \backslash \{qr\}$
8: If $ans(qr, \mathcal{A})$ is null, go to step7, else return false.

In what follows, two lemmas and the main theorem are presented, which shows the conversion of the integrity constraint validation into rewritten conjunctive queries evaluation over ABox.

Lemma 1. *Let $\mathcal{K}$ be a DL-Lite$_R$ knowledge base and Q a union of conjunctive queries over $\mathcal{K}$, then $ans(Q, \mathcal{K}) = \bigcup_{q_i \in Q} ans(q_i, \mathcal{K})$.*

Lemma 2. *Let $\mathcal{T}$ be the set of standard axioms expressed with DL-Lite$_R$, q a conjunctive query over $\mathcal{T}$, and PRQ the union of conjunctive queries obtained by the query rewriting. For every ABox $\mathcal{A}$ expressed with DL-Lite$_R$, $ans(q, <\mathcal{T}, \mathcal{A}>) = ans(PRQ, \mathcal{A})$.*

Proofs of above lemmas are similar to proofs in Ref. [11]. Based on Theorem 1 and above two lemmas, we conclude the correctness of Algorithm 1 in the following theorem.

Theorem 2. *Let $<\mathcal{K}, \mathcal{IC}>$ be an IC-KB, where $\mathcal{K} = <\mathcal{T}, \mathcal{A}>$, for every IC-axiom $\alpha \in \mathcal{IC}$, $\mathcal{K} \models_{IC} \alpha$ if and only if the algorithm IC_SATISFY($\mathcal{T}, \mathcal{A}, \alpha$) return true.*

Proof. If $\mathcal{K} \models_{IC} \alpha$, then from Theorem 1 we can see that $ans(\pi(\alpha), <\mathcal{T}, \mathcal{A}>) = \varnothing$. Further, from Lemma 1 we know that each answer of union of conjunctive queries over $\mathcal{K}$ is the same as the union of answers of each conjunctive query over $\mathcal{K}$. Thus, for each query $q \in \pi(\alpha)$, $ans(q, <\mathcal{T}, \mathcal{A}>) = \varnothing$ holds. Moreover, according to Lemma 2, it is obvious that $ans(PRQ, \mathcal{A}) = \varnothing$. Thus, for each query $q \in PRQ$, $ans(q, \mathcal{A}) = \varnothing$ holds. It means either each query $q \in \pi(\alpha)$, $ans(q, \mathcal{A})$ is null, or there exist some query $q' \in Q'$ such that $ans(q', \mathcal{A})$ is not null, and answers of all rewritten queries for q' over $\mathcal{A}$ is null, where $Q' \subseteq Q$. For the first case, the Algorithm 1 may iteratively go to step 3 to step 6, and proceed in step 3 with $\mathcal{IC}$ is null, and return true. For the second case, we only consider q'. Since for each rewritten query $rq \in Qr$, where Qr represent the set of rewritten queries for rq, $ans(rq, \mathcal{A})$ is null, the Algorithm 1 may go to step 5, and continue to check the following query in CQ. Thus, in this case, it may also proceed in step 3, and return true. Vice versa.

4   Evaluation

4.1   Instance Evaluation

In the following, we compare the Tao's work refer in [9] with our validation. The main difference between the two methods is whether TBox axioms are contained in standard *DL-Lite_R* knowledge bases. Without TBox axioms, validations of both methods can directly use SPARQL queries to check the satisfaction. Whereas, Tao's method may lose results of standard reasoning when TBox axioms related with constraints are in *DL-Lite_R* knowledge bases, which may lead to wrong results. The following example is given to demonstrate the desirable behavior.

Example 1. *Consider a DL-Lite_R knowledge base $\mathcal{K}$ that consists of the ABox $\mathcal{A}_5$: {GraduateStudent(John), Professor(Mary), teachTo(Mary, John)}, TBox $\mathcal{T}$ contains {GraduateStudent $\sqsubseteq$ Student} and $\mathcal{IC}$ contains ax_4 : Professor $\sqsubseteq$ $\exists teachTo.Student$.*

The IC-axiom ax_4 is translated into a union of conjunctive query q_7.
q_7 : $Professor(x) \wedge (\mathbf{not}\ teachTo(\mathbf{x}, \mathbf{y}) \vee \mathbf{not}\ Student(\mathbf{y}))$.
Eliminate the disjunction, obtain conjunctive queries
q_8 : $Professor(x) \wedge \mathbf{not}\ teachTo(\mathbf{x}, \mathbf{y})$.
q_9 : $Professor(x) \wedge \mathbf{not}\ Student(\mathbf{y})$.
According to Tao's approach, the query evaluation over ABox $\mathcal{A}_4$ is directly used. $ans(q_9, \mathcal{A}_4) = \{Mary\}$, $\mathcal{K}$ violate the IC-axiom ax_3, the violated individual is Mary.

However, in fact, according to standard axiom GraduateStudent $\sqsubseteq$ Student and the assertion GraduateStudent(John) we can infer Student(John), and ax_4

should be satisfied. When using our approach, it is necessary to rewrite q_9, since the right-side of TBox axiom "GraduateStudent $\sqsubseteq$ Student" and q_9 has the same concept "Student". The rewritten query is as follows:

q_{10} : Professor$(x) \wedge$ **not** GraduateStudent$(\mathbf{y})$.

$ans(q_8, \mathcal{A}_2) = \varnothing$ and $ans(q_{10}, \mathcal{A}_2) = \varnothing$. Therefore, $\mathcal{K}$ satisfies the IC-axiom ax_4, we obtain more intuitive results for integrity constraint validation.

Table 2. Details of data sets

Datatype	ABox	#ic(sat\uns))	#s	#cq	#prq
Onto1	155	20(14/6)	3	25	7
Onto2	155	20(13/7)	9	25	10
Onto3	1196	40(28/12)	3	46	25
Onto4	8487	42(29/13)	3	48	37

4.2   Experiments

Experiments were performed on a PC running Windows 7 OS with Intel core i5 3.20 GHZ CPU and 8 GB RAM. Our experiments have been executed on the well-known Lehigh University Benchmark (LUBM) ontology benchmark [15]. We use the file "Univ-Bench" in LUBM as standard axioms and modify the expressivity to DL $DL\text{-}Lite_R$ according to experiment requirements. Moreover, several axioms are added as IC-axioms expressed with DL $\mathcal{SROIQ}$ according to user requirements. Our experiments were performed on data sets automatically generated by "Univ-Bench".

We then check the satisfaction of integrity constraints based on the consistency checking [16], which also added negation as failure into the integrity constraint validation, and the rewriting w.r.t. different scalability of ontology. The information about these data sets is given in Table 2. For each data set, five aspects are listed: size of dataset(abox), the number of IC-axioms in IC, which contains the number of satisfied and violated axiom(#ic(sat/uns)), the number of standard axioms in $\mathcal{T}$(#s), the number of conjunctive queries translated from IC-axioms(#cq), the number of rewritten conjunctive queries by query rewriting(#prq).

Figure 1 indicates the efficiency gap between rewriting based validation (Onto$_i$_Rewrite) and consistency based validation(Onto$_i$_Cons). The horizontal axis represents different IC-axioms and the vertical axis represents the validation times with different dataset. For the rewriting based validation, validation times just exceed 100 ms for each ontology, whereas validation times of consistency checking validation far exceed 1000 ms and 10000 ms for Onto3 and Onto4. Thus, we can get the following result: with the increase of ontology scale, the efficiency gap is clearer.

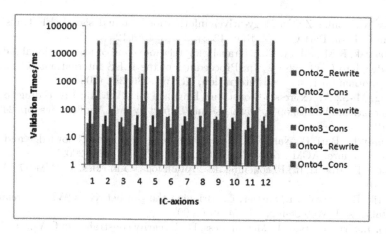

Fig. 1. Validation time w.r.t. different scale of datasets (Color figure online)

5   Conclusion

Motivated by the correctness of ontology data in social computing applications, the integrity constraint validation for *DL-Lite_R* ontology is proposed. Based on the translation of integrity constraint axioms to conjunctive queries, the integrity constraint validation is converted into the conjunctive query answering. Furthermore, by reformulating IC-axioms and corresponding conjunctive queries, standard axioms are no longer dependent on the integrity constraint satisfaction and will be discarded. Therefore, the integrity constraint validation is essentially reduced to the query evaluation over database instances, and can sufficiently use the query optimism in relational database management systems. Moreover, as the result of validation performance shows that the rewriting based algorithm is superior to the traditional query answering and optimize the integrity constraint validation. By doing this, users are more convenient to identify data in complex social computing applications.

Acknowledgements. This work was supported in part by NSFC under Grant Nos. 61502199, 61402196, 61272208; the Opening Foundation of Key Laboratory of Symbolic Computation and Knowledge Engineering of Ministry of Education under Grant No. 93K172016k02; the Independent Fund of Dalian Minzu University under Grant No. DC201501060.

References

1. Antoniou, G., Harmelen, F.: Web Ontology Language-OWL. In: Staab, S., Studer, R. (eds.) Handbook on Ontologies. International Handbooks on Information Systems, pp. 67–92. Springer, Heidelberg (2004)
2. Parameswaran, M., Whinston, A.B.: Social computing: an overview. Commun. Assoc. Inf. Syst. (2007)

3. Petr, K., Kouba, Z.: Ontology-driven information system design. IEEE Trans. Syst. Man Cybern. Part C Appl. Rev. **42**(3), 334–344 (2012)
4. Albarrak, K.M., Sibley, E.H.: Translating relational and object-relational database models into OWL models. In: Proceedings of the IEEE International Conference on Information Reuse and Integration, IRI, pp. 336–341 (2009)
5. Fang, M.-S.: An expressive constraint language for OWL. In: Proceedings of 23rd International Workshop on Database and Expert Sytems Applications, pp. 249–253 (2012)
6. Donini, F., Bari, P., Nardi, D.: Description logics of minimal knowledge and negation as failure. ACM Trans. Comput. Logics **3**(2), 177–225 (2002)
7. Motik, B., Rosati, R.: Reconciling description logics and rules. J. ACM **57**(5), 1–62 (2010)
8. Motik, B., Horroks, I., Sattler, U.: Bridging the gap between OWL and relational databases. J. Web Semant. **7**, 74–89 (2009)
9. Tao, J., Sirin, E., Bao, J., McGuinness, D.: Integrity constraints in OWL. In: AAAI, pp. 11–15 (2010)
10. Baader, F., Calvanese, D., McGuinness, D., et al.: The Description Logic Handbook: Theory, Implementation and Application, 2nd edn. Cambridge University Press, Cambridge (2007)
11. Calvanese, D., Giacomo, D.G., Lembo, D., Lenzerini, M., Rosati, R.: Tractable reasoning and efficient query answering in description logics: the DL-Lite family. J. Autom. Reasoning **39**, 385–429 (2007)
12. Glimm, B., Horrocks, I., Lutz, C., Sattler, U.: Conjunctive query answering for the description logic SHIQ. In: IJCAI, pp. 399–404 (2007)
13. Horrocks, I., Kutz, O., Sattler, U.: The even more irresistible SROIQ. In: Proceedings of the 10th International Conference on Principle of Knowledge Representation and Reasoning, pp. 57–67 (2006)
14. Calvanese, D., Eiter, T., Ortiz, M.: Answering regular path queries in expressive description logics: an automata-theoretic approach. In: Proceedings of the Twenty Second AAAI Conference on Artificial Intelligence, pp. 391–396 (2007)
15. Guo, Y., Pan, Z., Heflin, J.: LUBM: a benchmark for OWL knowledge base systems. J. Web Semant. **3**(2), 158–182 (2005)
16. Ouyang, D., Cui, X., Ye, Y.: Integrity constraint in OWL Ontology based on grounded circumscription. Front. Comput. Sci. **7**(6), 812–821 (2013)

Research and Implementation of Single Sign-on in Enterprise Systems Application Integration

Zhihong Wang[1,4], Yi Guo[1,2(✉)], Wenwu Tang[1], Yongbin Xu[1],
Bicheng Feng[3], and Qin Hou[3]

[1] Department of Computer Science and Engineering,
East China University of Science and Technology, Shanghai 200237, China
yguo1110@ecust.edu.cn
[2] School of Information Science and Technology, Shihezi University,
Shihezi 832003, China
[3] Shanghai Huateng Software Systems Co., Ltd., Shanghai, China
[4] Shanghai HiKnowledge Information Technology Co., Ltd., Shanghai, China

Abstract. Single Sign-on (SSO) is an effective unified authentication and authorization mechanism that makes users access to all integrated applications that trust each other from one single-sign-on site. This paper starts from the basic principle analysis of CAS, a single sign-on solution. In order to solve the problems of complex permissions assignment and inflexibility of mounting existent applications, an extended CAS SSO solution, named Ext-CAS SSO, is proposed in a procedure of enterprise systems integration. Furthermore, this paper states the full structure, working protocol and concrete implementation of Ext-CAS SSO in different scenarios. At last, a questionnaire survey regarding the related systems is conducted to testify the user experience.

Keywords: SSO · CAS · Enterprise systems application integration · Unified authentication

1 Introduction

With the development of enterprise information building, there are more and more business systems, such as project management system, office system and mailing system, have been applied in enterprises. Labor-intensive and numeric-intensive tasks are substituted with computers to promote working efficiency and other subside benefits. However, "Information Islands" [1] gradually appear in this procedure. Under this isolated island, each system has its own specific authentication mode and database which holds pairs of user name and password. Since each user are obliged to register with private information when they visit the resources, come up following problems,

This work is supported by the National Natural Science Foundation of China (Grant number 61462073).

such as, (1) multiple accounts increase maintenance costs and decline maneuverability; (2) repeated logins increase the possibility of credential breach; (3) overwhelming obstacle in information sharing [2, 3]. Under this situation, a technology of Single Sign-on emerges.

The single sign-on (SSO) [4, 5] systems help users tackle those problems and enable users to access to all mutual-trusted application systems with a universal login module. At present there are several commercial and open source solutions for SSO such as IBM Tivoli SSO, Sun SSO and CAS SSO etc. This paper mainly focus on CAS SSO developed by Yale University. CAS is developed as an independent web application, which can be deployed in Tomcat as a service. Moreover, CAS supports multiple programming language clients, each of those clients is feasible in SSO development with CAS. However, there are also some problems existing in enterprise systems application integration based on CAS, such as complex permissions assignment, inflexibility of mounting existent applications and so on. In order to deal with those problems, an extended CAS SSO is proposed in diverse scenarios with different strategies. (1) Strategy of Acegi with CAS. (2) Strategy of JDBC with CAS. (3) Strategy of Web Service. The implementation details will be described in the following sections.

The rest of this paper is organized as follows. Section 2 introduces the basic principle of SSO and analyzes the problems of enterprise systems application integration with CAS. Section 3 proposes the extended CAS SSO (Ext-CAS SSO) solution and implements SSO in diverse scenarios based on Ext-CAS. Furthermore, a questionnaire survey is conducted to testify the user experience of the related systems. Section 4 concludes this paper and provides future research works.

2   Fundamental Principle of CAS

2.1   System of CAS Institutions

In terms of CAS framework, there are two essential parts in CAS, including CAS Server and CAS Client.

2.1.1   CAS Server

CAS Server [4] is a platform-independent web application which can be deployed in any application server, such as Tomcat, Jetty etc. It is primarily responsible for user authentication and application authorization management. Besides, CAS server can handle user credentials such as username and password through reading database or XML files.

2.1.2   CAS Client

CAS client [4] is deployed together with web servers which need SSO protection. It is in charge of intercepting users' requests for the protected systems and redirecting them to the CAS Server. In addition, CAS Client supports many web environments such as Java, .Net, PHP, Perl, Ruby etc.

2.2   CAS Essential Protocol

The CAS framework protocol is shown in Fig. 1.

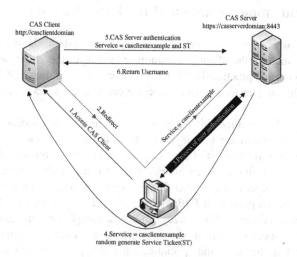

Fig. 1. Basic protocol of CAS

With the aim to guard the SSO protection resources, CAS Client filters each user request and parses it to check whether Service Ticket (ST) contained or not. Without ST, the request must be redirected to CAS Server to authenticate user login credentials. In the certification procedure, CAS Server generates a random ST which would be returned to CAS Client. Lastly, a trusty channel is built between CAS Client and CAS Server. Then, the CAS Client will service the current user who has the correct ST.

2.3   Problems in the Application of CAS

Through analyzing the principle of CAS protocol, several problems have been found in enterprise systems application integration, such as:

(1) Difficulty of assigning complex permissions in CAS Server. In the CAS-based application system integration, there is only one overall user table maintained in CAS Server. However, the server cannot well complete assigning complex permissions for each business system. So the permission assignment tasks should be done by every business system after logon. But the existent CAS is not well developed about it.

(2) Inflexibility of mounting existent applications in CAS Client. A variety of configuration information should be modified when a new application system is integrated into the SSO system. Take java client for instance, in order to realize SSO based on CAS, web.xml file should add some filters, e.g. CAS Authentication Filter, CAS Validation Filter, HttpServletRequest Filter etc., and some parameters such as casServerLoginUrl, serverName, casServerUrlPrefix etc. It is obvious that the configuration information is not only cumbersome but also may affect system's stability.

Moreover, there are also several constraints during the practical configuration, such as the existent applications lack unity and coherence.

3 Design and Implementation of the Ext-CAS SSO

In the process of enterprise system integration, SSO system should involve many diverse factors, such as the client programming language, the client architecture etc. This paper proposes a practicable SSO solution which combines CAS with several strategies in different scenarios to overcome the original CAS drawbacks described in Sect. 2.3.

Without loss of generality, this paper adopts different rights allocation policies for the mainstream programming language clients including Java and .Net. On the one hand, the permission assignment of Java clients is relatively complicated because Java is open source. Thus, this research adopts Acegi to deal with complex permissions assignment. On the other hand, as .Net framework has strong integration, the complex permissions can be assigned with JDBC. And for other programming language clients, this paper also adopts basic JDBC strategy to assign complex permissions. In addition, there are some special applications in enterprise system integration, such as higher-privileged applications and special-demanded applications. However, the original CAS has nothing to do with them. In this research, Web Service is adopted to integrate them into the SSO system. As previously stated, this paper proposes an extended CAS SSO (Ext-CAS SSO) framework for general applications and special applications. The implementation of Ext-CAS SSO framework is shown in Fig. 2.

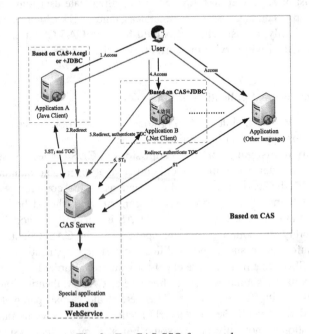

Fig. 2. Ext-CAS SSO framework

Papers not complying with the LNCS style will be reformatted. This can lead to an increase in the overall number of pages. We would therefore urge you not to squash your paper.

3.1 Strategy of Acegi with CAS

Acegi (Spring Security) [7] is a Security Framework based on Spring IOC (Inversion of Control) and AOP (Aspect Oriented Programming) which provides permission management for enterprise applications. It is one the best security framework available for the Java platform. So this security framework is adopted in this paper to deal with the complex permissions. Under this circumstance, Acegi would assign complex permissions with its own strategy of casAuthoritiesPopulator, which can obtain user and role information from RDBMS/LDAP after CAS authentication. In this way, the drawbacks of complex permissions assignment can be solved well in Java clients.

3.2 Strategy of JDBC with CAS

Java Data Base Connectivity (JDBC) [8] is a Java API for programming language that defines how a client may access a database. It provides several methods for querying and updating data in a database. And it also provides a benchmark that can builds advanced tools and interfaces which allows database developers to program database applications. Thus this research adopts JDBC to deal with the complex permission assignment for .Net and other programming language clients, such as Perl, PHP etc. It is obvious that CAS Server is used for user authentication and complex permissions is assigned by the clients. So the drawbacks of complex permissions assignment can also be well solved.

3.3 Strategy of WebService

In consideration of the existent applications in enterprise system integration, there are always with complex and unclear framework. And sometimes there are also existing some higher-permission applications. For the above mentioned applications, it is difficult to integrate them into CAS SSO system directly, which may affect the stability and security of the whole systems.

On the premise of ensuring stability and security, the special applications can attach themselves to the generic applications to realize SSO based on Web Service. In this case, the special login evidence, consists of Service Ticket and Username, would be verified through Web Service interface. If validation passes, user can log into the special applications without repeated logins. It is obvious that to SSO for special applications through Web Service doesn't change any configuration files of original systems and also has no effect on stability and security of the whole systems.

3.4   Implementation of Ext-CAS SSO

Contraposed the generic and special applications, this paper proposes several different strategies to solve the problems of complex permissions assignment and inflexibility of mounting existent applications. Moreover, this paper also implements Ext-CAS-based SSO in diverse scenarios for different demands.

3.4.1   Scenario One

The Single Sign-on systems allow users to log into several applications with only one authentication. In scenario one, there are several different programming language applications without special applications. The implementation of Ext-CAS SSO framework is shown in Fig. 3.

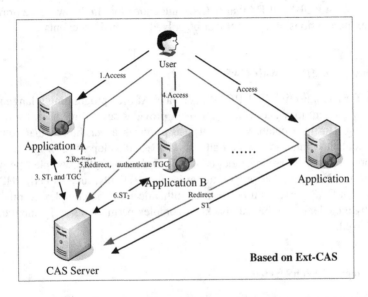

Fig. 3. The scenario one of Ext-CAS SSO

The workflow of Ext-CAS SSO in scenario one is described as follows:

(1) An unauthenticated user tries to access application A which is under the Ext-CAS protection (User can access to any one of Ext-CAS protected applications, this paper makes application A for example).
(2) Application A could not find user's information in the session, and then redirects the request to CAS Server to validate.
(3) If validation is successful, CAS Server generates Ticket-Granting Ticket (TGT) [9] and Service Ticket (ST1), and return ST1 to the application. Then the application redirects the request to CAS Server to verify the legitimacy of ST1. After verification successful, CAS Server writes Ticket-Granting Cookie (TGC, cookie stored TGT) [9] in the browser.

(4) When the user accesses to other application, such as application B, the application redirects the request to the CAS Server for login confirmation with TGC.

(5) After confirmation successful, CAS Server generates a new ST2 for application B and returns it to application. Then the application redirects the request to CAS Server to verify the legitimacy of ST2. After verification successful, the application B allows user to access its service.

(6) For the other applications, the step is the same as (4) and (5).

3.4.2 Scenario Two

Apart from the generic applications, described in scenario one, it also has several special applications in enterprise system integration, such as higher-privileged applications and special-demanded applications. The original CAS SSO cannot integrate them into SSO systems well. Nevertheless, the WebService-based Ext-CAS SSO can treat them well, and the framework is shown in Fig. 4.

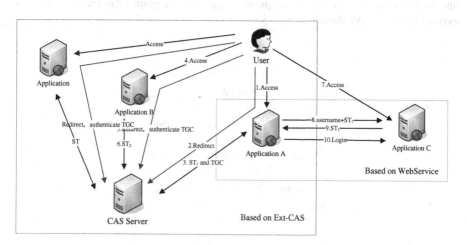

Fig. 4. The scenario two of Ext-CAS SSO

The Ext-CAS-based authentication process in scenario two is the same as scenario one Sect. (3.4.1). And the Web-Service-based authentication process is as follows:

(1) An unauthenticated user tries to access application A (or any one of Ext-CAS protected applications).

(2) Application A could not find user's information in the session, and then redirects the request to CAS Server to validate.

(3) If validation is successful, CAS Server generates Ticket-Granting Ticket (TGT) and Service Ticket (ST1), and return ST1 to the application. Then the application redirects the request to CAS Server to verify the legitimacy of ST1. After verification successful, CAS Server writes Ticket-Granting Cookie (TGC, cookie stored TGT) in the browser.

(4) User tries to access the special application C which is attached to application A.

(5) Because application A and C is connected by Web Service, the username and ST1 are sent to application C through browser.
(6) After obtaining ST1, the application C calls Web Service interface of application A to verify the legitimacy of ST1.
(7) If verification is successful, the application C allows user to access its service.

3.5   User Experience Analysis

3.5.1   User Experience Quantitative Criteria of Ext-CAS

According to the ISO definition [10], User experience (UX/UE) [11–13] is a person's perceptions and responses that result from the use or anticipated use of a product, system or service. In addition, the ISO also points out that the UX can be quantified by some factors.

Peter Morville [14] is known as the "founding father" of information architecture and he proposes several factors to quantify the UX and graphs the factors with honeycombed figure. As shown in Fig. 5.

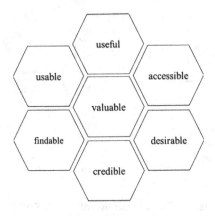

Fig. 5. The honeycombed graph of UX factors

Moreover, Robert Rubinoff [15] believes that the UX is based on the attitude of customer first and stresses the utility and esthetics of a product, system or service. Consequently, he separates web's UX into four aspects: Branding, Usability, Function and Content, as shown in Fig. 6. These aspects are always used to quantify and evaluate web's UX.

Against the background of practical application and investigation, the UX quantitative criteria of Ext-CAS (Fig. 7) are given in this paper, which combines Peter's criteria with Robert's criteria.

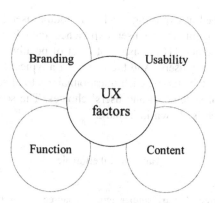

Fig. 6. Web UX factors

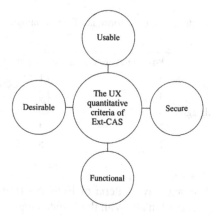

Fig. 7. The UX quantitative criteria of Ext-CAS

Where:

Usable is that whether user can utilize a product or service to finish their tasks.
Desirable refers to whether a product or service can meet user's emotional needs.
Secure means that user's privacy can be well protected by the product or service.
Functional is that a product or service can offer them convenience for their tasks.

3.5.2   Experiment Design of User Experience

In order to obtain users' experience data, the most direct way is to ask users through a self-report method. And the data types of self-report are divided into subjective and objective. But for the research of user experience, subjective data obtained from users are objective.

In this research aimed at exploring the login module user experience, we adopt a questionnaire survey to collect the user experience data based on a Likert Scale method. Likert Scale method [16] consists of a set of positive or negative statements that need to be answered by users. And the format of a typical five-level Likert item is shown in Table 1. After the survey, each questionnaire's total score will be calculated based on the polarity of statements and users' choices. The scoring formula of questionnaire in this research is shown in Table 2.

Table 1. Likert Scale

Likert Scale				
Strongly agree	Agree	Neither agree nor disagree	Disagree	Strongly disagree

Table 2. Scoring formula of questionnaire

Scoring formula		
Likert Scale	Positive statement/score	Negative statement/score
Strongly agree	5	1
Agree	4	2
Neither agree nor disagree	3	3
Disagree	2	4
Strongly disagree	1	5

According to the UX quantitative criteria of Ext-CAS (Fig. 7), we redact several statements for each aspect and 12 in all. With those statements, a UX survey is done for Ext-CAS SSO system, CAS SSO system and original system. The UX factors and testing statements are shown in Table 3 and the questionnaire is shown in Appendix A.

Table 3. UX factors and testing statements

Usable	Q1: Convenient to log on system
	Q2: Fast to log on system
	Q3: Systematic integral capability is stability
Desirable	Q4: System interface design is simple and attractive
	Q5: System structure is chaos
	Q6: Difficult to grasp the complex operations of system
Security	Q7: My personal information is protected by the system
	Q8: The sign design of system is security
Functional	Q9: System has satisfied my demand
	Q10: System makes it easy to finish my job
	Q11: A lot of time has been saved by system
	Q12: System makes my job more effective

3.5.3 UX Experimental Result

In this paper, an anonymous questionnaire is conducted among 150 experimental subjects selected from system users. There are 144 (96.0 %) samples return. After removing the invalid questionnaires, there are 139 (92.7 %) samples remaining, which satisfy the sample research of statistical requirements. In addition, there are totally 150 users in the survey, of whom 54.0 % are male and 46.0 % are female. And according to the report of China Internet Network Information Center (CNNIC) on January 22, 2016 [17], the proportion of male internet user is 53.6 %. That is to say, the samples selected are reasonable.

After the survey, the effective total score of each UX factor for Ext-CAS SSO, CAS SSO and original login system will be collected separately. Then we'll calculate the final score of each factor by a five-point scale formula (Eq. 1) and graph the result with relative plot (Fig. 8).

$$\text{Final Scores} = \text{User Choices Scores} / \text{Indicator Total Scores} * 5 \qquad (1)$$

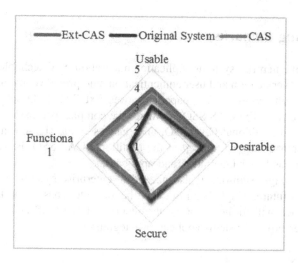

Fig. 8. The each score comparison of UX factors (Color figure online)

In Fig. 8, it is obvious that almost each UX factor of Ext-CAS is superior to the original system and CAS besides "Secure". For this phenomenon, we can suspect that the shorter service time of Ext-CAS leading to users' misjudgment are probably the main causes of the original system's "Secure" score are slight higher than Ext-CAS. But the total score of four aspects of Ext-CAS is obviously higher than others, as is shown in Fig. 8. Generally speaking, the user experience of Ext-CAS SSO has made a remarkable improvement, and outmatches the original system and CAS SSO (Fig. 9).

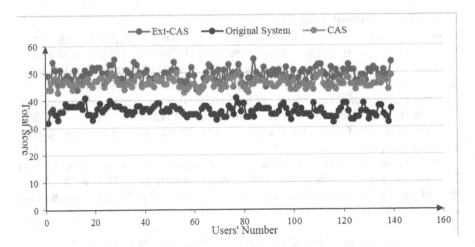

Fig. 9. The total score of UX (Color figure online)

4   Concluding Remarks

In the process of enterprise systems application integration, SSO technology can realize uniform identity verification and user authentication, and provide convenience methods to log on the multi-systems. This paper proposes Ext-CAS SSO solution based on CAS SSO framework. Ext-CAS SSO optimizes the complex permission assignment of multiple clients in traditional CAS SSO, and realizes single sign-on in practical scenarios for different needs. Compared to the traditional CAS SSO, Ext-CAS SSO this paper proposed has better integration and application.

In addition, single sign-out is also needed in enterprise systems application integration. In the future, single sign-out can be designed based on Ext-CAS SSO framework, which will further improve the security of Ext-CAS and expand its usefulness in the enterprise systems application integration.

Appendix A: Questionnaire

Dear consumer,

In order to improve the design and functionality of Ext-CAS SSO system, we are now engaged in the survey of system's UE. Sincerely thank you for sparing time from your busyness to answer our questionnaire. Thank you for your cooperation and assistance.

Please answer questions in the tables and choose the correct one in the corresponding circle according to your real situation about the Ext-CAS, CAS and the original.

The questionnaire is conducted by a secret way and all information will be kept in strict confidence.

Part One: Please answer the following personal information.

Gender	O Male	O Female
Frequency of system utilization	O Everyday	O 1 ~ 3 times a week
	O 1 ~ 3 times a month	O Occasional

Part Two: Please choose the correct one in the corresponding circle according to your real feels about the Ext-CAS (A), CAS (B) and the original system (C).

	Strongly agree			Agree			Neither agree nor disagree			Disagree			Strongly disagree		
	A	B	C	A	B	C	A	B	C	A	B	C	A	B	C
Q1: Convenient to log on system.	O	O	O	O	O	O	O	O	O	O	O	O	O	O	O
Q2: Fast to log on system.	O	O	O	O	O	O	O	O	O	O	O	O	O	O	O
Q3: Systematic integral capability is stability.	O	O	O	O	O	O	O	O	O	O	O	O	O	O	O
Q4: System interface design is simple and attractive	O	O	O	O	O	O	O	O	O	O	O	O	O	O	O
Q5: System structure is chaos.	O	O	O	O	O	O	O	O	O	O	O	O	O	O	O
Q6: Difficult to grasp the complex operations of system.	O	O	O	O	O	O	O	O	O	O	O	O	O	O	O
Q7: My personal information is protected by the system.	O	O	O	O	O	O	O	O	O	O	O	O	O	O	O
Q8: The sign design of system is security.	O	O	O	O	O	O	O	O	O	O	O	O	O	O	O
Q9: System has satisfied my demand.	O	O	O	O	O	O	O	O	O	O	O	O	O	O	O
Q10: System makes it easy to finish my job.	O	O	O	O	O	O	O	O	O	O	O	O	O	O	O
Q11: A lot of time has been saved by system.	O	O	O	O	O	O	O	O	O	O	O	O	O	O	O
Q12: System makes my job more effective.	O	O	O	O	O	O	O	O	O	O	O	O	O	O	O

References

1. Fu, X.: Enterprise Information Integration Management: Theory and Case. Beijing University of Posts and Telecommunications Press, Beijing (2006)
2. Hope, P., Zhang, X.: Examining user satisfaction with single sign-on and computer application roaming within emergency departments. Health Inform. J. **21**(2), 107–119 (2015)
3. Sun, Q., Jiang, B., Xin, Y.: Research and implementation of single sign–on for multi-technology based on web services. J. Changzhou Univ. **24**(1), 62–64 (2012)

4. Zhang, Y.L., Huang, H.: Design and realization of the component of single sign-on based on web services and session verification. In: Applied Mechanics and Materials, vol. 411, pp. 481–485 (2013)
5. Højgaard, J., Kiniry, J.R.: Securing single sign-on systems with executable models. Masters thesis, IT University of Copenhagen (2013)
6. Li, Y., Du, L., Fan, C., et al.: Design and implementation of a CAS single sign-on authentication service based on LDAP. Adv. Comput. Sci. Technol. 65, 243 (2014)
7. Wang, Y.: The design and implementation of universal user authority management system based on acegi. Softw. 34(7), 46–50 (2013)
8. Ganesh, S.G., Sharma, T.: Building database applications with JDBC. In: Oracle Certified Professional Java SE 7 Programmer Exams 1Z0-804 and 1Z0-805, pp. 281-315. Apress (2013)
9. Luo, S.: Agile Acegi and CAS: Building Secure Java Systems. Electronic Industry Press, Beijing (2007)
10. ISO FDIS 9241-210:2009. Ergonomics of human system interaction - Part 210: Human-centered design for interactive systems (formerly known as 13407). International Organization for Standardization (ISO). jithin dev
11. Karapanos, E.: User experience over time. In: Karapanos, E. (ed.) Modeling Users' Experiences with Interactive Systems. SCI, vol. 436, pp. 57–83. Springer, Heidelberg (2013)
12. Hassenzahl, M.: User experience (UX): towards an experiential perspective on product quality. In: Proceedings of the 20th International Conference of the Association Francophone d'Interaction Homme-Machine, pp. 11–15. ACM (2008)
13. Hassenzahl, M.: User experience and experience design. In: Soegaard, M., Dam, R.F. (eds.) The Encyclopedia of Human-Computer Interaction, vol. 2. The interaction design foundation, Aarhus (2013)
14. Morville, P.: User experience design. Ann Arbor: Semantic Studios LLC (2004)
15. Rubinoff, R.: How to quantify the user experience, 5(10) (2008). Accessed 2004
16. Likert, R.: A technique for the measurement of attitudes. Archives of psychology (1932)
17. https://www.cnnic.net.cn/hlwfzyj/hlwxzbg/201601/P020160122469130059846.pdf. Statistical Report on Internet Development 37th China Internet Network (2016)

Research on SVM Plant Leaf Identification Method Based on CSA

Xuhui Zhang, Yang Liu[✉], Haijun Lin, and Yukun Liu

Harbin University of Science and Technology, Harbin, China
{150080,475281216}@qq.com

Abstract. In view of the longer training and recognition time of plant leaf classifier, this paper proposes a method of blade recognition based on the combination of clonal selection algorithm and support vector machine. The method uses the blade geometry and texture features as the identification feature, building a blade recognition classifier based on support vector machine, in order to obtain the optimal kernel function parameter and the penalty factor, using cross validation characteristics of immune clonal selection algorithm to optimize the kernel function parameter and the penalty factor. Experimental results show that compared with BP neural network and other two methods, the proposed method has a higher recognition accuracy and training speed.

Keywords: Image identifying · Support vector machine · Clonal selection algorithm · Geometric features · Texture features

1 Introduction

The identification and classification of plant is mainly used for the distinguish of floristics, genetic relationship between plants and evolution rule of exploration research direction. Using the detailed data of agricultural plants to seed and nursery has very important significance for the improvement of agricultural production [1]. In the field of plant identification, a lot of kinds of classification methods has been developed, in which recognition method based on plant leaves has been an important research direction of this field. Compared with the traditional classification method of leaves, plant leaves recognition based on image analysis has the advantages of long survival time, high work efficiency and widely data processing [2].

At present,the domestic and foreign researchers mainly use the leaf color, shape and other characteristics to identify and classify plant. Application of image recognition technology in feature selection, algorithm performance and classification has made great progress [3]. In 2009, Chinese Academy Hefei Institutes of Physical Science presented a robust supervised manifold learning algorithms classify plant leaves, reduces the computational complexity of the algorithm to identify; In 2010, Singh and others used the SVM method based on the binary tree to identify and classify 32 plant

Science and technology research projects of education department in Heilongjiang (12541126).

W. Che et al. (Eds.): ICYCSEE 2016, Part II, CCIS 624, pp. 171–179, 2016.
DOI: 10.1007/978-981-10-2098-8_20

leaves [4]; In 2011, Rossatto used volume fractal dimension and Naive Bayes classi-fication to identify leaf images; And domestic researchers also made a series of pro-gress in the use of shape, texture features to identify the leaves [5]; 2013 Zhang utilizing the methods of combing CSA and K near to establish a database blades including 100 kinds of plant leaves and 16 samples, described a method which can improve the recognition rate in the conditions of a small training size and character-istics [6]; In 2015, Ding et al. proposed a method of more characteristic of plant leaves image recognition based on the difference value between D-LLE algorithm which effectively improves the recognition accuracy of the blade [7]. However, there still exist the problem of the characteristic parameter acquisition complexity, training data capacity, long training time [8–10].

This paper proposes a method combining CSA and SVM for identifying blade, regional blade geometry and texture features as the main identification features, while improving blade identification process classifier, and the blades of the image classifi-cation information processing by the classifier, determining the type of plant, reducing the training time and increasing leaf identification accuracy.

2 Leaf Feature Extraction

Feature extraction is to get the effective features from the leaf image after image preprocessing, which identifies according to certain strategy extraction. Because the edge position is very important for the extraction of the shape feature of the blade, the filter should be paid attention to keep the details of the edge part clear enough. Under certain conditions, the median filter does not have the details of the image produced by the linear filtering method, and it can reduce the random noise points. Therefore, the paper uses median filter to deal with the noise of the blade. For plant leaf images under simple background, the segmentation based on threshold is usually selected. After the comparison of several threshold segmentation algorithms, the OTSU method is selected as the segmentation method. Firstly, the occurrence probability of each gray level is calculated based on the histogram, and the gray level is divided into two categories by the threshold variable K, then the class variance of each class is obtained, and select the K value making the class variance max as the optimal threshold value. The pretreat-ment results of median filter and threshold segmentation are shown in Fig. 1.

(a)Original leaf image   (b) Leaf image after median filtering (c) Leaf image after segmentation

Fig. 1. Leaf image after preprocessing

2.1 Geometric Features

Geometric features extraction in the recognition leaf has been widely used, which the first to be classified as a standard feature vector, and therefore the research on extraction methods relatively mature. This paper, using the poplar blade as a sample, calculates 10 values of geometric characteristics of the blade, and the results are shown in Table 1.

Table 1. Poplar blade geometry eigenvalues

Feature names	Eigenvalues	Feature names	Eigenvalues
Axis aspect ratio	1.3569	Roundness	0.6042
Rectangle degree	0.8007	Eccentricity ratio	1.4095
Convex area ratio	0.9477	Shape parameter	0.6361
Ratio of the circumference irregularities	1.0232	Circumference radius ratio	2.7618
Spherical property	0.6123	Lobation	0.6487

2.2 Texture Features

Texture is one of the inherent characteristics of the surface or structure, which is an important feature to observe and identify objects. It can be used to identify and classify the blade surface color or grayscale regular distribution. This paper uses fractal dimension, energy, moment of inertia, local stationary, entropy these 5 characteristics. The paper uses the poplar blade as a sample and the texture feature values are shown in Table 2.

Table 2. Poplar blade Texture eigenvalues

Feature names	Eigenvalues
Fractal dimension	1.9684
Energy	0.8262
Moment of inertia	0.8355
Local stationary	0.8395
entropy	0.6873

3 Leaf Recognition Classifier Design

We need classify the image by the classifier after the pretreatment and extraction of the blade image feature value. This paper studies a kind of SVM (Support Vector Machine) classifier based on the CSA (Clonal Selection Algorithm) to extract the blade geometric features and texture features of initialization parameters. It constitutes characteristic vectors and processes the data of the blade using the method of combining CSA and SVM, then trains to form the mature antibody library of the leaf, and completes the classification and identification of leaf samples.

3.1 SVM

The SVM is a kind of common machine learning method, which can seek the best compromise between the complexity of the model and the learning ability according to the limited sample information. The SVM can automatically search for a better classification ability to distinguish between support vector, thereby constituting classifier can maximize the interval between class and class.

Provided training samples $X = \{x_1, x_2, \ldots x_n\} \subset R^D$, y_i is the category of x_i, $y_i \in \{-1, +1\}$, $i = 1, 2, \ldots n$.

The linear discriminant function for D dimensional space is

$$f(x) = w \cdot x + b \tag{1}$$

Classified hyper plane equation

$$w \cdot x + b = 0 \tag{2}$$

Among them, w is the normal vector of classified hyper plane, b is the classification threshold. The classification interval of classified hyper plane is $2/\|w\|$, minimizing the $\|w\|$ makes the classification interval maximum, classification of hyper plane will be the correct classification of the two categories.

In the case of linear non division, it is necessary to map the samples in low dimensional space to high dimension space, and then find the optimal classification hyper plane which can separate the sample points in high dimension space, so there are

$$y_i(w \cdot \phi(x_i) + b) - 1 \geq 0 \tag{3}$$

Form the relevant theory of functional, we can learn when a kernel function $K(x_i, x_i)$ satisfies the Mercer condition, the kernel would then correspond to a inner product in the transformation space, which is:

$$K(x_i, x_j) = \phi(x_i) \cdot \phi(x_j) \tag{4}$$

Therefore, the choice of a suitable kernel function can be applied to the optimal classification hyper plane. The optimal classification function is:

$$f(x) = \text{sgn}(\sum_{i=1}^{n} a_i^* y_i K(x_i, x) + b^*) \tag{5}$$

3.2 CSA

The basic theory of CSA is: When a specific antigen invades the body and internally selected immune cells to clone which can identify and eliminate the corresponding antigen, making it activated, differentiated and proliferated, then conducting an immune response to eliminate antigen-presenting cells. The characteristic data of the treated leaves are used as the antigen, and the formula of the antigen initialization is shown in the formula (6).

$$A_g = \left(b_1(A_g), b_2(A_g), L, b_i(A_g)L, b_N(A_g) : G\right) \tag{6}$$

Among them, $b_i(A_g)$ is the i-th characteristic value of the antigen, G is the corresponding classification categories for this antigen. The part of the blade and the random data value are used as antibodies, antibody initialization formula is:

$$A_g = \left(b_1(A_g), b_2(A_g), L, b_i(A_g)L, b_N(A_b) : B\right) \tag{7}$$

Among them, $b_i(A_b)$ is the i-th characteristic value of the antibody, B is retention value which is antigen corresponding category of its highest affinity at the mature antibody repertoire.

The affinity between antibody and antigen is a standard to measure the matching of two, Affinity formula is shown below.

$$Af(A_g, A_b) = \frac{1}{1 + dis(A_g, A_b)} \tag{8}$$

When the immune system is activated, antibody starts to clone antibody proliferation, which improves the global convergence of the efficiency, and the affinity is higher so that the clone scale of antibodies is larger. The clone scale factor is determined by the formula (9):

$$g_m = \text{int}\left[m \cdot \left(Af_i / \sum_{i=1}^{N} Af_i\right) \cdot ha_i\right] \tag{9}$$

Among them, g_m is clone scale, Af_i is the i-th antigen affinity, ha_i represents the minimum hamming distance of the i-th antigen and antibody, m is the setting value related to the clone scale, m is 12 in this paper.

3.3   CSA and SVM Combination

Using SVM to classify plant leaves data when data for SVM classification of plant leaves optimal search problem RBF kernel function kernel parameters γ and the penalty factor value c. To try using CSA to select the kernel function parameters of SVM model. This article will define antigen as the target function, randomly generated population of antibodies as a core parameter solutions, the affinity between antigens and antibodies regard as the approximate extent feasible solution with the optimal solution, using SVM K-fold cross-validation method are represented. The step flow chart of SVM based on CSA are shown in Fig. 2. Firstly, normalized leaf characteristics data, according to the training data of blade, initializes the algorithm, including antibody scale parameters, optimal number of individuals, high frequency of mutation rate, maximum algebra or the highest affinity threshold, antigen SVM fold cross validation of the k values. Then the antibody binary decoding algorithm is set up, and the binary initial antibody group A_b is randomly generated. Each antibody of the antibody group is the candidate solution of the optimization problem. Through binary decoding, the K-fold cross-validation method is used to calculate the affinity between antibody and antigen. According to the affinity value, m optimal antibodies P_m are selected from the A_b antibody group to get the binary clone population A_b^c. Mutate the binary clone population A_b^c high frequently to obtain the variation of antibody population $A_b^{c^*}$. It is affected by the double restriction of mutation rate and affinity. And A_b population excludes the low affinity to get A_b^l, then the rest clone population recalculates affinity of antibody group A_b^* after mutation, selects the affinity of the largest m antibodies, adds to the Ab^n to get A_b^l, and Ab^n instead of A_b. Finally, according to the

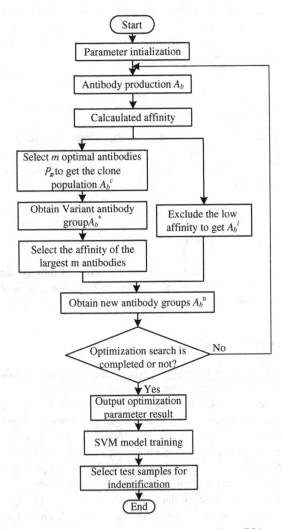

Fig. 2. Flowcharts of SVM trainer based on CSA

obtained optimized RBF kernel function and the parameter C and γ, the blade characteristic SVM model is trained, and the test samples are selected to identify the blade.

4   Results

In this paper,we select 120 kinds of plant leaves from Harbin Erlong Mountain forestry and Harbin botanical garden as experimental subjects and collect 40 samples of each leaves. The collected image of the blade is reduced to 800 × 600 pixels for preprocessing. Among these leaves, we extract 20 of them as the training samples and other 20 as testing samples. Then analyze the training time, recognition time and recognition rate of different methods. The system interface of SVM plant leaves recognition system based on clonal selection algorithm is shown in Fig. 3.

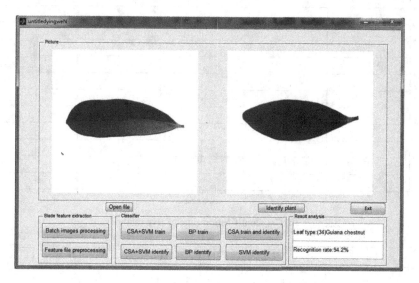

Fig. 3. SVM plant leaves identification system interface based on CSA

Experimental results show that for the same number of blades sample, training time, recognition speed and recognition rate of SVM classifier based on CSA, BP neural network classifier [11], CSA classifier, SVM classifier and the method combing CSA and K Nearest Neighbor these five methods are compared. This paper takes poplar leaves for example to test 20 times. Comparative results are shown in Fig. 4.

Experimental results are showed in Table 3.

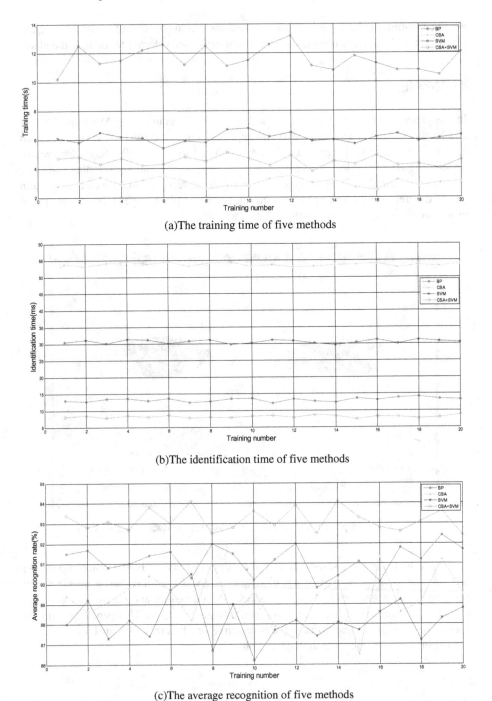

(a)The training time of five methods

(b)The identification time of five methods

(c)The average recognition of five methods

Fig. 4. The comparative results of five methods for poplar leaves

Table 3. Comparative value results of training, recognition and average recognition rate

Recognition method	Training time/s	Recognition time/ms	Average recognition rate/%
BP neural network	11.8	13.2	91.3
CSA	3.1	54.3	89.1
SVM	6.3	30.6	88.6
CSA + KNN	3.2	8.7	91.4
CSA + SVM	4.6	8.2	93.1

5 Conclusion

In view of the long training time of plant leaf identification method, this paper has studied the identification method about plant leaf based on CSA and SVM. Firstly, using the method of median filtering and threshold segmentation processes image, and extract feature of that leaf. And then, by using the method of CSA and SVM, the sample training is carried out, and the recognition of the classifier is also carried out. It can be seen from the experimental results that the method of this paper can shorten the training and recognition time, and improve the recognition rate, which indicates that the method is feasible.

References

1. Lei, G., Wang, S., Huang, H., Jiang, L., Huang, Z.: Plant growth control system based on internet of things. J. Anhui Agric. Sci. **42**, 7662–7664 (2014)
2. Chen, Y., Zhou, P.: Research on shape and texture feature extraction of piant leaf images. J. Zhejiang Sci-Tech Univ. **30**, 394–398 (2013)
3. Lin, C., Yuhuan, S., Licheng, J.: Application of immune clone selection algorithm to image segmentation. J. Electron. Inf. Technol. **28**, 1169–1173 (2006)
4. Singh, K., Gupta, I., Gupta, S.: SVM-BDT PNN and fourier moment technique for classification of leaf shape. Int. J. Sig. Process. **3**, 67–68 (2010)
5. Rossatto, D.R., Casanova, D., Kolb, R.M., et al.: Fractal analysis of leaf-texture properties as a tool for taxonomic and identification purposes: a case study with species from neotropical melastomataceae (mi-conieae tribe). Plant Syst. Evol. **291**, 103–116 (2011)
6. Zhang, N., Liu, W.: Plant leaf recognition method based on clonal selection algorithm and K nearest neighbor. J. Comput. Appl. **33**, 2009–2013 (2013)
7. Ding, J., Dong, L., Yan, Q.: Recognition method of multi-feature plant leaves based on D-LLE algorithm. Comput. Eng. Appl. **51**, 158–163 (2015)
8. Deng, X., Jiao, L., Yang, S., Wu, Q.: Color image segmentation in a multidimensional space based on clonal selection algorithm. J. Electron. Inf. Technol. **32**, 1792–1796 (2010)
9. He, P., Huang, L.: Feature extraction and recognition of plant leaf. J. Agric. Mechanization Res. **58**, 168–170 (2008)
10. Neto, J.C., Meyer, G.E., Jones, D.D.: Plant species identification using elliptic fourier leaf shape analysis. Comput. Electron. Agric. **50**, 121–134 (2006)
11. Hou, T., Yao, L., Kan, J.: Plant recognition research based on shape features of leaf. Hunan Agric. Sci. 4, 123-125,129 (2009)

Research on Technology of Twin Image Recognition Based on the Multi-feature Fusion

Yanqing Wang[1(✉)], Yipu Wang[1], Chaoxia Shi[2], and Hui Shi[3]

[1] School of Computer Science and Technology,
Harbin University of Science and Technology, Harbin 150080, China
wyq0325@126.com, 575626476@qq.com
[2] School of Computer Science and Engineering,
Nanjing University of Science and Technology, Nanjing 210094, China
stonexia@sohu.com
[3] Henan Electric Power Transmission and Transformation
Engineering Company, Zhengzhou 450051, China
772563659@qq.com

Abstract. In order to improve the accuracy and stability of fruit and vegetable image recognition by single feature, this project proposed multi-feature fusion algorithms and SVM classification algorithms. This project not only introduces the Reproducing Kernel Hilbert space to improve the multi-feature compatibility and improve multi-feature fusion algorithm, but also introduces TPS transformation model in SVM classifier to improve the classification accuracy, real-time and robustness of integration feature. By using multi-feature fusion algorithms and SVM classification algorithms, experimental results show that we can recognize the common fruit and vegetable images efficiently and accurately.

Keywords: Face recognition · Twins image recognition · Attribute extraction · Feature combination · SVM

1 Introduction

Analyzing and extracting the information of 2D or 3D facial images are the most important areas of pattern analysis. Kinship verification by facial pattern is a new problem in pattern analysis and computer vision. It can be applied in many fields, such as history and genealogy research, database management, forensic and finding the missing family members and so on. Automatic kinship validation has great potential value, but it is a new field and there are very few research results. This paper only researches the technology of twin's image recognition. Xia et al. [1] assume that the similarity of face images between children and children's parents is very high. The experiments used extended transfer subspace learning (TSL) to get the best discriminated subspace from the target class (including children and their parents) by an intermediate class (including parents' youth images). Guo and Wang [2] calculate the probability of two individual's characteristics by the familial characteristics of the training set's hallmark (for example, eyes, nose and mouth). The accuracy of dataset under uncontrolled illumination conditions is over 75 %, so it can be seen kinship

© Springer Science+Business Media Singapore 2016
W. Che et al. (Eds.): ICYCSEE 2016, Part II, CCIS 624, pp. 180–187, 2016.
DOI: 10.1007/978-981-10-2098-8_21

verification through facial image is feasible. This paper selects a variety of different facial attributes as features to recognize twin's images.

2 Twins Image Recognition Process

This paper achieves twin's kinship recognition. In the experiments, each subject has two images (a positive side or a half image, a profile image) under uncontrolled illumination conditions. The experiment consists of three data sets, the positive image dataset, the profile image dataset and full image dataset. The experiment has 120 group pictures. 92 group pictures of them [3] were got from network, and the remaining 28 group pictures were got from the author and author's friends.

This section describes the various steps of twin's image recognition algorithm, and it's shown in Fig. 1.

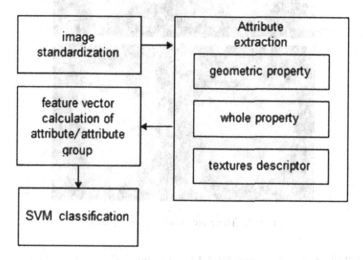

Fig. 1. It's the flowchart of twin's image recognition, and it shows the overall process.

1. Standardizing all the pictures in the dataset.
2. Extracting the follow features and forming all kinds of feature vectors. For each attribute of the same subject, the feature vectors can be obtained successively.
3. Establishing a data set by selecting two subjects (namely a and b). The data set is composed of an equal number of positive samples (twins image pairs) and negative samples (non-twin image pairs) randomly. Feature vector $v^{(ab)}$ is a Euclidean distance of a pair of subjects a and b. In n-dimensional space, the element in the feature vector $v^{(ab)}$ equal the attribute. The feature vector of a pair of subjects is exchangeable, namely $v^{(ab)} = v^{(ba)}$.
4. We should select the appropriate kernel function and parameters to build SVM classifier for each attribute or attribute combination in the data set. We can choose the most relevant characteristic variables based on the result of SVM.

2.1   Image Standardization

In the pretreatment step, we will adjust the image size for reunification. We chose reference points in the same position of positive side and profile side images. Then we standardized images geometrically [4]. As shown in Fig. 2, circular points are reference points and square points are corresponding points.

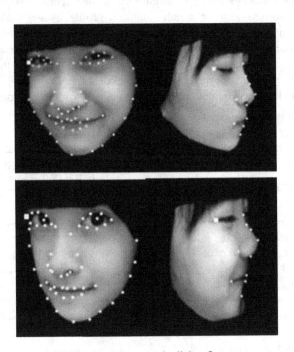

Fig. 2. There are standardizing figures.

2.2   Attribute Extraction and Facial Expression

Geometric property

Attribute 1 Coordinate. The property is defined as A_{COO}. It contains the position of the face normalized coordinate (x,y). Coordinate (x,y) is a standardized coordinate, so that the same type images have the same position.

Attribute 2 Special face triangle. The experiment focuses on the property and the property is defined as A_{TRI}. We chose the important reference points and connect them to form triangles with important facial features. The triangle is called special face triangle, as shown in Fig. 3. The data of special face triangle are sub data of the attribute, such as the length of side of triangles, the angle of triangles and the relative position of triangles. Since the images are different and the coordinate position of the special face triangle also are different, so it is not easy to compare the absolute coordinate of special face triangle. Thus, the coordinates used in this experiment are average of coordinates after standardization. The data of special face triangle such as divided length can effectively capture the shape of a spherical face and directly or indirectly describe some key facial dimensions (for example the distance between mouth and eyes, mouth width, etc.).

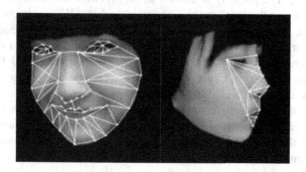

Fig. 3. This figure shows the attribute of special face triangle.

Whole property

Attribute 3 PCA. The property is defined as A_{PCA}. Eigen face dimension reduction technique [5] has been widely used in the field of face image analysis. Before principal component analysis (PCA) is applied to a set of available samples, each image used in the experiment first is cut into the standard area. After dropping as much as possible interference information (such as background pixels), we can conduct PCA transform and select several feature dimensions that represents the most image information. These dimensions should total represent more than 95 % of the amount of information. Then we should conduct PCA dematrixing and get new features.

Setting the face image size is m* n, and obtained the column vector after the transformation is $\{x_i\}, x_i \in R^{m*n}, i = 1, \ldots, N$. The estimated mean vector is $\mu = \frac{1}{N} \sum_{i=1}^{N} x_i$. The covariance matrix is $S_T = \sum_{i=1}^{N} (x_i - \mu)(x_i - \mu)^T = XX^T$. The dimensionality reduction formula of projection matrix A which takes the first k eigenvectors of the largest eigenvalues is $y = (X - \mu)A_k$.

Textures descriptor property

Attribute 4 RIC-LBP. The property is defined as A_{LBP}. The rotation-Invariant co-oc-currence of adjacent LBPs (RIC-LBP) presented by foreign researchers is an extension of the classic local binary patterns (LBPs) [6]. The algorithm is designed to improve the illumination invariance and rotation invariance. In a preliminary test of the experiment, RIC-LBP better than other LBPs extensions.

We can get a series of initial LBP value by continuously rotating circular neighborhood when we calculated the property. The lowest value is the LBP value of the neighborhood. The formula is $LBP_{P,R}^{ri} = \min\{ROR(LBP_{P,R}, i)|i = 0, 1, \ldots, P-1\}$, and it mean the p-digit X mobile I times.

Attribute 5 SIFT. The property is defined as A_{SIFT}. Scale invariant feature transform (SIFT) [7, 8] is a popular local image descriptors. In fact, SIFT has become a computer vision standard, widely used in the field of object recognition, image registration, content-based image retrieval and other fields. Therefore SIFT-based descriptor has been used to characterize the nearby area of image coordinates. SIFT can conduct unified constant scaling and rotation on a monochrome image, as well as some invariant affine distortion and illumination changes. Recently, researchers presented different types of descriptors, so we can apply different SIFT algorithms for different color schemes. The method can effectively improve the lighting invariance and discrimination of the image. All colors descriptors of the process is calculated as a separate assessment SIFT descriptors, and its size is 128. We can associate the corresponding color model of each channel to a simple vector.

2.3 SVM-Based Classification

The experiment used support vector machine (SVM) to build a classifier. SVM is a perfect machine learning techniques that has been proven successful in many applications [9, 10]. The sample space is mapped into a high dimensional feature space by SVM classifiers through a nonlinear mapping. So that nonlinear separable problem of sample space is translated into linearly separable problems of feature space. Since the establishment linear SVM learning machine in a high-dimensional feature space, using SVM almost no increase in computational complexity compared with the linear model to avoid the "curse of dimensionality". The core of SVM is radial basis function in this experiment and the radial basis function is $K(x, y) = \exp\left(\left(\frac{-|x-y|^2}{d}\right)^2\right)$.

The experiment optimized the kernel parameters by fivefold cross-validation techniques and grid search of parameters space [9].

3 Results and Discussion

3.1 Recognition Accuracy

In the experiment, we defined property groups, which are geometry property group A_{GEO}, textures property group A_{GRA} and whole property group A_{ALL}. We calculated the classification accuracy for each property and property group. Geometry property group contains coordinate property and special face triangle property, namely $A_{GEO} = \{A_{COO} \cup A_{TRI}\}$. Texture property group contains RIC-LBP property and SIFT

Table 1. Properties and property groups recognition accuracy table

Property/property groups	A positive side image accuracy (%)	A profile side image accuracy (%)	Image accuracy (%)
Coordinate (A_{COO})	54.5	59.2	65.7
Special face triangle (A_{TRI})	66.6	69.6	74.2
Geometry (A_{GEO})	68.1	75.1	82.1
PCA (A_{PCA})	64.2	68.2	72.6
RIC-LBP (A_{LBP})	61.7	59.7	64.1
SIFT (A_{SIFT})	70.5	67.8	77.5
Textures (A_{GRA})	73.8	77.3	84.9
All properties (A_{ALL})	74.2	78.4	87.2

property, namely $A_{GRA} = \{A_{LBP} \cup A_{SIFT}\}$. Whole property group contains all of properties, namely $A_{ALL} = \{A_{GEO} \cup A_{PCA} \cup A_{GRA}\}$. Table 1 shows the properties and property groups recognition accuracy.

By analyzing Table 1, the following conclusions can be drawn.

1. If we get more information, we can obtain higher classification accuracy.
2. In the geometric attributes analysis, coordinate position itself does not provide meaningful results.
3. SIFT as texture descriptors obtained more accurate results than the geometric properties, which shows SIFT more suitable for capturing twin's features. On the contrary, in most cases, RIC-LBP has poor performance.
4. In single attribute experiment, SIFT attribute gets the best average results.
5. Property groups obtained better results than single property in the group. Whole property group received the highest classification accuracy. In other words, we should try to get heterogeneous information as more as possible, the accuracy will be better.

3.2 Recognition Time

The experiment running time is an important standard to measure an algorithm. Table 2 shows the running time of properties and property groups. The running time is the time to calculate and get result in the experiment. From the data in the table, the calculation and classification time of special face triangle property, RIC-LBP property and SIFT property are similar. The experiment used the shortest time is 12 ms and the longest time is 27 ms. The recognition accuracy is over 80 %.

It can be seen that the recognition time of algorithm is within the acceptable range.

Table 2. Properties and property groups recognition time table

Property/property groups	A positive side image time (ms)	A profile side image time (ms)	Image time (ms)
Coordinate (A_{COO})	2	2	4
Special face triangle (A_{TRI})	3	4	7
Geometry (A_{GEO})	6	6	12
PCA (A_{PCA})	1	1	2
RIC-LBP (A_{LBP})	3	4	7
SIFT (A_{SIFT})	3	4	6
Textures (A_{GRA})	6	7	13
All properties (A_{ALL})	12	15	27

4 Conclusion

Twin's images classification results show that combining different kind of feature properties can effectively achieve high accuracy. The highest precision can reach 87.2 % in the experiment. In the experiment, special face triangle property's test result is higher than conventional PCA property's and RIC-LBP property's test result, so that the special face triangle property is more suitable for kinship recognition. The experiment can expand to recognition other kinship direction in the future, for example, parents and children, parents and grandchildren and so on. In addition, we used the whole face image to recognition twins now. In the future we can try to use part of face image to identify whether having kinship. We should build larger twin image database for future experiments. A sufficient amount of data could enhance the accuracy of the experimental results.

Acknowledgments. This paper has been supported by the National Natural Science Foundation of China (Grant No. 61371040).

References

1. Xia, S., Shao, M., Fu, Y.: Kinship verification through transfer learning. In: Proceedings of the Twenty-Second International Joint Conference on Artificial Intelligence. Springer, Spain (2011)
2. Guo, G., Wang, X.: Kinship measurement on salient facial features. In: Proceedings of IEEE Transactions on Instrumentation & Measurement. springer (2012)
3. China Taiwan Network. http://www.taiwan.cn
4. Mliki, H., Fendri, E., Hammami, M.: Face recognition through different facial expressions. J. J. Sig. Process. Syst. **81**, 1–14 (2015)
5. Lu, J., Liong, V.E., Wang, G.: Joint feature learning for face recognition. IEEE Trans. Inf. Forensics Secur. **10**(7), 1 (2015)
6. Cai, J., Chen, J., Liang, X.: Single-sample face recognition based on intra-class differences in a variation model. Sensors **115**(1), 1071–1087 (2015)

7. Patrik, Š., David, S.: Progress in SIFT-MS: breath analysis and other applications. Mass Spectrom. Rev. **30**(2), 236–267 (2011)
8. Yanqing, W., Biao, L., Zhuang, L.: Applied technology in unstructured road detection with road environment based on SIFT-HARRIS. J. Adv. Mater. Res. **1014**, 259–262 (2014)
9. Shalev-Shwartz, S., Singer, Y., Srebro, N.: Pegasos: primal estimated sub-gradient solver for SVM. Math. Program. **127**(1), 3–30 (2011)
10. Dong, J.-X., Krzyżak, A., Suen, C.Y.: A Fast SVM Training Algorithm. In: Lee, S.-W., Verri, A. (eds.) SVM 2002. LNCS, vol. 2388, pp. 53–67. Springer, Heidelberg (2002)

Sliding Window Network Coding for Free Viewpoint Multimedia Streaming in MANETs

Chao Gui[1], Chengli Huang[2], Baolin Sun[1(✉)], and Xiaoyan Zhu[3]

[1] School of Information and Engineering, Hubei University of Economics, Wuhan 430205, China
gui_chao@126.com, blsun@163.com
[2] Computer Engineering Department, Guangdong Youth Vocational College,
Guangzhou 510507, China
Chengli_huang@163.com
[3] School of Mathematics and Computer Science, Jianghan University,
Wuhan 430056, China
zhuxy@jhun.edu.cn

Abstract. Network coding (NC) can prove to be particularly useful in multimedia streaming applications, where it could help to increase the network throughput and data persistence in Mobile Ad hoc NETwork (MANET). Sliding-window Network Coding is a variation of NC that is an addition to multimedia streaming and improves the data delay on MANETs. In this paper, we propose a Sliding Window Network Coding for free Viewpoint Multimedia streaming in MANETs (SWNC-VM). SWNC-VM preserves the degree distribution of the encoded packets through the recombination at the nodes. SWNC-VM enables to control the decoding complexity of each sliding-window independently from the packets received and recover the original data. The performance of the SWNC-VM is studied using NS2 and evaluated in terms of the encoding overhead, decoding delay, packet loss probability and throughput when a packet is transmitted. The simulations result shows that the SWNC-VM with our proposition can significantly improve the network throughput and encoding efficiency.

Keywords: MANET · Network coding · Free viewpoint multimedia · Encoding efficiency

1 Introduction

Mobile Ad hoc NETworks (MANETs) consist of a large number of mobile nodes which are randomly deployed with mobility, computation, and communication capabilities.

This work is supported by The National Natural Science Foundation of China (No. 61572012), The Key Natural Science Foundation of Hubei Province of China (No. 2014CFA055, 2013CFB309), and The Science Foundation of Ministry of Education of Wuhan of China (No. 2011060). A Project Funded by the Priority Academic Program Development of Jiangsu Higher Education Institution (PAPD). Jiangsu Collaborative Innovation Center on Atmospheric Environment and Equipment Technology (CICAEET).

W. Che et al. (Eds.): ICYCSEE 2016, Part II, CCIS 624, pp. 188–197, 2016.
DOI: 10.1007/978-981-10-2098-8_22

These types of networks have many advantages, such as self-reconfiguration and adaptability to highly variable mobile characteristics like the transmission conditions, wireless channel distribution characteristics and power level [1–5]. Although packet error can be tolerated to some extents in most multimedia streaming applications, excessive packet losses are unacceptable because it can lead to the degradation of quality of experience (QoE) to the end mobile users. In order to improve the reliability of multimedia streaming, many techniques and protocols have been developed.

Network coding (NC) is one of the recent breakthroughs in communications research. It has first been proposed in [2] and potentially impacts all areas of communications and networking. NC was in fact devised based on algebraic operations over the finite fields $GF(2^8)$, which yields to very complex multiplications from the computational and thus energetic perspective. The advantages of network coding come however at the price of additional computational complexity, mainly due to the packet encoding and decoding process. Ahlswede $et\ al.$ [2] considered network coding for solving energy consumption in MANETs. But they did not give a specific network coding implementation. Network coding is not new, it has its foundation in different systems which were highlighted by Yeung [5] while discussing the historical perspective of network coding, that lead to the seminal paper on network coding. Magli $et\ al.$ [6] reviews the recent work in NC for multimedia applications and focuses on the techniques that fill the gap between NC theory and practical applications. Yang $et\ al.$ [7] introduced R-Code, a network coding-based reliable broadcast protocol for Wireless Mesh Network (WMNs) that achieves 100 % packet delivery ratio with low transmission overhead and relatively short broadcast latency. Thomos $et\ al.$ [8] propose a scheme that builds on both rateless codes and network coding in order to improve the system throughput and the video quality at clients. Liu $et\ al.$ [9] propose a novel low-complexity distributed multiple description coding (LC-DMDC) method in order to further improve the error resilience of distributed video coding while maintaining low encoder complexity and good rate-distortion performance in hostile network conditions.

This paper proposes a Sliding Window Network Coding for free Viewpoint Multimedia streaming in MANETs (SWNC-VM). Our contributions towards sliding window, low-complexity, NC are in the following.

First, we provide a thorough description of sliding window and network coding for free viewpoint multimedia streaming in MANET (SWNC-VM), a novel class of network codes. SWNC-VMs achieve controlled encoding and decoding complexity thanks to the joint design of the encoding, decoding, and recombination processes.

Second, an analytical model of SWNC-VM decoding complexity is derived allowing to matching the decoding computational cost to the capacity of the mobile node.

Third, the performance of the SWNC-VM is studied using NS2 and experimentation to assess the encoding efficiency, the decoding complexity of SWNC-VM enabled mobile node.

The rest of the paper is organized as follows. Section 2 discusses the some related work. Section 3 describes models of sliding encoding window model in MANET. Some simulating results are provided in Sect. 4. Finally, the paper concludes in Sect. 5.

2 Related Works

Mobile multi-hop live streaming for free viewpoint multimedia has been extensively studied, with some focusing on minimizing delay [6–9], while others studying energy optimization [10–13]. There has also been work on optimizing the overlay structure or a multi-radio multi-channel network [14–16]. Our work differs by studying a novel wireless live free viewpoint multimedia streaming network with collaborative anchor pulling and NC generation, and its joint optimization problem.

Qin *et al.* [10] proposed an energy-saving scheme for wireless sensor networks based on network coding and duty-cycle (NCDES). The scheme determines the node's status based on the ID information which embedded in data information. When combining network coding and duty-cycle in wireless sensor networks, it will reduce transmission coding coefficients and retransmissions. Jiang *et al.* [11] proposes an energy-efficient multicast routing approach to achieve the data forwarding in the multi-hop wireless network. Analysis of the multi-hop networks energy metric and energy efficiency metric. Then the corresponding models are given network coding is used to improve network throughput. Xie *et al.* [12] propose an approximation algorithm, which is called ADCMCST (Algorithm with the minimum number of child nodes when the depth is restricted), to construct a tree network for homogeneous wireless sensor network, so as to reduce and balance the payload of each node, and consequently prolong the network lifetime. Antonopoulos *et al.* [13] proposed a network coding-based cooperative ARQ (NCCARQ) scheme for wireless networks. Compared to simple cooperative ARQ protocols, the proposed solution improves up to 80 % the energy efficiency of the system without compromising the offered QoS in terms of throughput and delay.

The work in [14], both consider a general cooperative recovery scenario for full recovery, but has not considered the joint problem of packet pulling and sharing. They employ Instantly Decodable Network Coding with XOR operation, which only considers the decodability of individual NC packets. Zhang *et al.* [15] study a novel live free viewpoint video streaming network where each user pulls a subset of anchors from the server via a primary channel. To enhance anchor availability at each user, a user generates NC packets using some of its anchors and broadcasts them to its direct neighbors via a secondary channel. Bayat *et al.* [16] proposed a peer-to-peer (P2P) framework for the deployment of live video streaming applications over P2P overlay networks. The proposed framework provides support for flash-crowds, decentralizes decision making and makes use of network coding to reduce bandwidth consumption. Ostovari *et al.* [17] purposed a lightweight triangular inter-layer NC instead of the general form of inter-layer NC, to reduce the time complexity of the optimization.

Although some network coding algorithms are proposed to improve network performance, most of these approaches do not consider mobile multimedia streaming scenario. In our work, we target mobile multimedia streaming networking problem in MANETs. We take into account maximizing network throughput and consider minimizing decoding delay. By constructing the appropriate network coding structure, we can achieve the higher free viewpoint multimedia streaming.

3 Sliding Encoding Window Model

We now focus to the sliding encoding window and the random network coding approaches. When using this approach, not all packets need to be coded together in a generation, just the ones in the same window. When using this approach, not all packets need to be coded together in a generation, only the code group operating in the same window. This simplifies the solving of the Gaussian–elimination on the receiver node, but requires constant feedback between the nodes to determine which packets have been seen at the receiver and thus remove them from the sender's linear combinations.

3.1 Encoding at the Source Node

The sliding encoding window model is a generalization of the insertion-only streaming model in which we seek to compute function f over only the most recent elements of the stream.

The network is represented as $G = (V, E)$ where V represents the set of nodes in the network and E denotes the set of directed edges. Each link $e = (i, j) \in E$ means that node i can transmit to node j. We assume links are symmetric that if $(i, j) \in E; (j, i) \in E$ as well. Whether two links interfere with each other depends on the interference model adopted.

The random linear network coding scheme [18] adopts a block transmission strategy which can approach the capacity with less feedback overhead. SWNC-VM adopts the encoding strategy similar to RLNC, but the block of packets to be encoded in each slot is sliding window forward at a constant speed V. For each packet P to transmit, the source node selects the blocks $(x_0, x_1, \ldots, x_{N-1})$ and coding vector $(c_0, c_1, \ldots, c_{N-1})$ to combine with in a sliding encoding window of size $1 \leq W \leq N$, The elements of the encoding vector that do not belong to the encoding window are equal to zero. The size of the sliding window $W = e - f + 1$, for N elements, there are $N - W + 1$ possible sliding windows of size W. A sliding encoding window of size W is a sequence of blocks $(x_f, \ldots, x_e)$ where $0 \leq f$, $e \leq N - 1$ and $f \leq 1$ and $e - f + 1 = W$. We define f_i and e_i the leading window and the trailing window of the i-th sliding encoding window. Figure 1 shows the sliding encoding window of SWNC-VM.

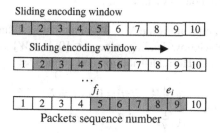

Fig. 1. Sliding encoding window of SWNC-VM.

Sliding window coding algorithm
Initialize $k = 1$, $i = 1$, and $n = R/(e-f)$
for each $k \in P$ do
    if $i < n$ then
        Transmit packet p_k *and sliding encoding window*
        $i \leftarrow i + 1$
    else
        Transmit $(x_f, \dots, x_e, c_0, c_1, \dots, c_{N-1})$
        $i \leftarrow 1$

3.2 Decoding at the Destination Node

After overhearing the coded symbols from the source, the receiver attempts to decode the original packets through Gauss elimination approach. A typical example of the decoding process is shown in Fig. 2, in which the Gauss-Jordan Elimination can be performed progressively as the coded symbol arrives and finally the original packets can be retrieved when the reduced matrix has full rank.

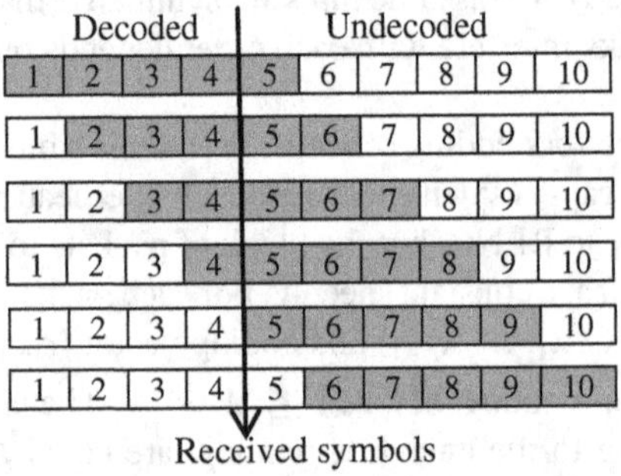

Fig. 2. Decoding vector of SWNC-VM.

Decoding Algorithm of SWNC-VM
The destination node receive $(x_f, \dots, x_e, c_0, c_1, \dots, c_{N-1})$
$\{x_0, x_1, \dots, x_{N-1}\} = $ Decode $(x_f, \dots, x_e, c_0, c_1, \dots, c_{N-1})$
To restore the original data $(x_0, x_1, \dots, x_{N-1})$

The encoder selects one out of the $N - W + 1$ available sliding encoding windows, randomly drawing the window leading edge $f \in [0, N - W]$ according to a discrete distribution function called Horn Distribution and defined as:

$$HD_W(f) = \begin{cases} \frac{W+1}{2N}, & \text{if } f = 0 \text{ or } f = N - W \\ \frac{1}{N}, & \text{if } 0 < f < N - W \end{cases} \tag{1}$$

Once f has been drawn, the trailing edge of the sliding encoding window is calculated as $e = f + W - 1$.

In this paper, we use a linear network coding scheme. The linear network coding scheme is an encoding method such that coding vector $c_i = (c_{i0}, c_{i1}, ..., c_{iN-1})$ is given, and input packet $X = (x_0, x_1, ..., x_{N-1})$ is converted into output packet P_i by the following expression.

$$P_i = \sum_{j=0}^{N-1} c_{ij} x_j \tag{2}$$

Then, the elements c_i of the encoding vector g are set to one with probability $p = 1/2$ for $i \in [f, e]$, with probability $p = 0$ otherwise. The destination node can decode input packets because the coding vector $c_i = (c_{i0}, c_{i1}, ..., c_{iN-1})$ and output packet data $P = (P_0, P_1, ..., P_{N-1})$ are obtained from the received packets, and an inverse matrix exists in G.

4 Simulation Experiments

We now focus to the sliding window and the random network coding approaches. When using this approach, not all packets need to be coded together in a generation, just the ones in the same window.

4.1 Simulation Scenario

In this section, we present various simulation results for the proposed a Sliding Window Network Coding for free Viewpoint Multimedia Streaming in MANETs (SWNC-VM) algorithm. We evaluate SWNC-VM in a free viewpoint multimedia streaming scenario where one source distributes a multimedia sequence to multiple cooperating receiver nodes. Nodes are randomly and uniformly located over a 1000 m × 1000 m area, with a node transmission range of 250 m [19]. Node mobility follows the random waypoint model. Link loss rate for anchor/NC packets is independent and identically distributed with mean 2 %. The multimedia stream is subdivided by the source as a sequence of generations of data with identical playout time C_t. Every C_t seconds, the source fetches a generation of multimedia and encodes and transmits packets for such generation only for the following C_t seconds. The generation distributed by the source is called seeding position within the multimedia stream. Due to its small size, beacon loss is assumed to be negligible (Table 1).

Table 1. Simulation parameters

Number of nodes	100
Network area	1000 m × 1000 m
Transmission range	250 m
Simulation time	600 s
Transmission range	250 m
Beacon period	100 ms
Communication model	Constant Bit Rate (CBR)
Message size (b_{msg})	512 bytes/packet

4.2 Simulation Results

We analyze the performance of SWNC-VM from the point of view of the encoding efficiency. The encoding efficiency of SWNC-VM depends on the generation size N and on the sliding encoding window size W, W range from 20 to 50.

Firstly, we test the SWNC-VM performance in encoding overhead. We plot the encoding overhead versus network size in Fig. 3. Figure 3 shows that the encoding overhead of SWNC-VM depends on sliding window W and hence on the average packet degree. SWNC-VM reduces the encoding overhead the better. The packet overhead reduces as the packet packets transmission increases because the number of nodes increases.

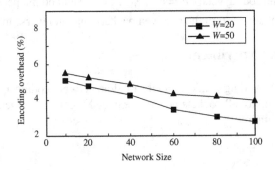

Fig. 3. Encoding overhead with different network size.

Then, we test the SWNC-VM performance in decoding delay. Figure 4 shows the impact of different encoding window density on the decoding delay. When the encoding window increases, the decoding delay increases. The Fig. 4 shows that SWNC-VM enable to change by adjusting sliding encoding window size W. Lower is better.

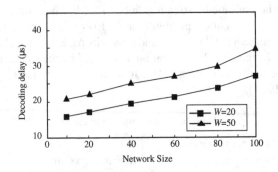

Fig. 4. Decoding delay with different network size.

We analyze the performance of SWNC-VM from the point of view of the packet loss probability. Figure 5 shows the actual processor load due to packet decoding as reported by the packet loss probability as a function of the encoding window size. The

Fig. 5 shows that SWNC-VM enable to increase by a factor of network size load by adjusting sliding encoding window size *W*. Lower is better.

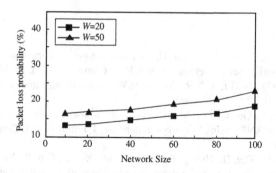

Fig. 5. Packet loss probability with different network size.

We analyze the performance of SWNC-VM from the point of view of the throughput. Figure 6 shows a comparison the sliding window network coding scheme in terms of throughput as a function of network size load by adjusting sliding encoding window size *W*. From the Fig. 6, it is clear that the SWNC-VM enable to increase by a factor of network size load by adjusting sliding encoding window size W. Higher is better.

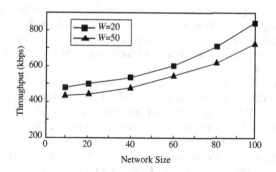

Fig. 6. Throughput with different network size.

5   Conclusion

This paper proposes a sliding window and network coding for free viewpoint multimedia streaming in MANET (SWNC-VM). First, we introduce sliding window, a class of network codes that preserves the packet degree distribution through the recombination at the coding nodes. Second, we design an analytical model of SWNC-VM decoding complexity is derived allowing to matching the decoding computational cost to the capacity of the mobile node. Third, the performance of the SWNC-VM is studied using NS2 and experimentation to assess the encoding efficiency, the decoding complexity of SWNC-VM enabled mobile node. The simulation result shows that SWNC-VM produce encoding overhead, decoding delay, packet loss probability and throughput. This technique can guarantee the same reliability while consume the least energy.

References

1. Sun, B.L., Gui, C., Song, Y., Chen, H.: A novel network coding and multi-path routing approach for wireless sensor network. Wirel. Pers. Commun. **77**(1), 87–99 (2014)
2. Ahlswede, R., Cai, N., Li, S.-Y.R., Yeung, R.W.: Network information flow. IEEE Trans. Inf. Theory **46**(4), 1204–1216 (2000)
3. Sun, B.L., Song, Y., Gui, C., Luo, M.: Network coding-based priority-packet scheduler multi-path routing in MANET using Fuzzy controllers. Int. J. Future Gener. Commun. Netw. **7**(2), 137–147 (2014)
4. Mohammed, A.H., Dai, B., Huang, B.X., Azhar, M., Xu, G., Qin, P., Yu, S.: A survey and tutorial of wireless relay network protocols based on network coding. J. Netw. Comput. Appl. **36**(2), 593–610 (2013)
5. Yeung, R.W.: Network coding: a historical perspective. IEEE Proc. **99**(3), 366–371 (2011)
6. Magli, E., Wang, M., Frossard, P., Markopoulou, A.: Network coding meets multimedia: a review. IEEE Trans. Multimedia **15**(5), 1195–1212 (2013)
7. Yang, Z., Li, M., Lou, W.: R-code: network coding-based reliable broadcast in wireless mesh networks. Ad Hoc Netw. **9**(5), 788–798 (2011)
8. Thomos, N., Frossard, P.: Network coding of rateless video in streaming overlays. IEEE Trans. Circ. Syst. Video Technol. **20**(12), 1834–1847 (2010)
9. Liu, W.H., Vijayanagar, K.R., Kim, J.: Low-complexity distributed multiple description coding for wireless video sensor networks. IET Wirel. Sens. Syst. **3**(3), 205–215 (2013)
10. Qin, T.F., Li, L.L., Yan, L., Xing, J., Meng, Y.F.: An energy-saving scheme for wireless sensor networks based on network coding and duty-cycle. J. Beijing Univ. Posts Telecommun. **37**(4), 83–87 (2014)
11. Jiang, D.D., Xu, Z.Z., Li, W.O., Chen, Z.H.: Network coding-based energy-efficient multicast routing algorithm for multi-hop wireless networks. J. Syst. Softw. **104**, 152–165 (2015)
12. Xie, S.D., Wang, Y.X.: Construction of tree network with limited delivery latency in homogeneous wireless sensor networks. Wirel. Pers. Commun. **78**(1), 231–246 (2014)
13. Antonopoulos, A., Verikoukis, C., Skianis, C., Akan, O.B.: Energy efficient network coding-based MAC for cooperative ARQ wireless networks. Ad Hoc Netw. **11**(1), 190–200 (2013)
14. Aboutorab, N., Sadeghi, P., Sorour, S.: Enabling a tradeoff between completion time and decoding delay in instantly decodable network coded systems. IEEE Trans. Commun. **62**(4), 1296–1309 (2014)
15. Zhang, B., Liu, Z., Chan, S.-H.G., Cheung, G.: Collaborative wireless freeview video streaming with network coding. IEEE Trans. Multimedia (2016). doi:10.1109/TMM. 2016.2518485
16. Bayat, N., Lutfiyya, H.: Network coding for coping with flash crowd in P2P multi-channel live video streaming. In: 11th International Conference on the Design of Reliable Communication Networks (DRCN), Kansas, KS, USA, 24–27 March 2015, pp. 243–246 (2015)
17. Ostovari, P., Wu, J., Khreishah, A., Shroff, N.B.: Scalable video streaming with helper nodes using random linear network coding. IEEE/ACM Trans. Netw. **24**, 1–14 (2015). doi:10.1109/TNET.2015.2427161
18. Swapna, B., Eryilmaz, A., Shroff, N.: Throughput-delay analysis of random linear network coding for wireless broadcasting. IEEE Trans. Inf. Theor. **59**(10), 6328–6341 (2013)

19. Shen, J., Tan, H.W., Wang, J., Wang, J.W., Lee, S.Y.: A novel routing protocol providing good transmission reliability in underwater sensor networks. J. Internet Technol. **16**(1), 171–178 (2015)
20. Waxman, B.: Routing of multipoint connections. IEEE J. Sel. Areas Commun. **6**(9), 1617–1622 (1988)
21. The Network Simulator - NS-2. http://www.isi.edu/nsnam/ns/

Demo Track

HierarSearch: Enhancing Performance of Search Engines by Mining Semantic Relationships Among Results

Qian Liu, Hongzhi Wang[✉], and Shaoying Song

Harbin Institute of Technology, Harbin, China
qliu10@ncsu.edu, wangzh@hit.edu.cn, 52jszdyjmm@sina.cn

Abstract. Web has a plethora of information. Modern search engines (such Google, Bing) can retrieve web pages based on keywords and return them in a ranking list. However, users with little knowledge in a certain domain would have difficulty in finding appropriate keywords to retrieve the web pages they intend to find. What's more, little interaction with search engine restrains users to do further investigation among the results. If users want to refine their queries, no assistance about potential related concepts/instances is provided. In this paper, we introduce a new search engine: HierarSearch, which enhances performance of search engines by providing an interactive keyword hierarchy generated from webpages retrieved. Users can interact with the tag cloud keyword hierarchy to know the semantic relationships among results, and also refine queries by selecting keywords in tag cloud. HierarSearch would generate refined keywords query "intelligently" with the help of a very large database, which understands human world.

Keywords: Information retrieval · Knowledge base · Data mining

1 Hierarchical Search

Web has a plethora of information. Modern search engines (such Google, Bing) can help users retrieve web pages and return them in a ranking list. However, inappropriate keywords generated by certain users: i.e. children, elder people, or people lack certain domain knowledge, would lead to almost irrelevant search results. What's more, if users who are unsatisfied with results would like to refine their queries, there is no other assistance is provided by search engines. The ranking list view restricts user's further investigation based on original keywords [4], since users neither know potential related concepts/instances, nor explore the relationships among results.

Such problems expose two shortcomings of common search engines. One is that general search engines are mostly based on keywords. However, the intention of human world is too complicated to be described in several keywords. Since keywords entered in search engines can have ambiguous meanings: i.e. the term "apple" may refer to either a fruit or a brand of computers. Also the same world entity can have different represen-tations: i.e. "Wei Wang", can also be represented as "W. Wang", and "W Wang". The other shortcoming is that the interaction between search engines and users is not enough. Search engines don't provide options to do further investigation except for a list of

© Springer Science+Business Media Singapore 2016
W. Che et al. (Eds.): ICYCSEE 2016, Part II, CCIS 624, pp. 201–205, 2016.
DOI: 10.1007/978-981-10-2098-8_23

ranked webpages. If users want to refine their results, they have to refine their keywords by their own. This raises the problem that sometimes although users know they need refined results, they don't know how to make it: i.e. a child wants to find interesting stories related to apple (fruit) when he types "apple story" in Google. However, he has no idea why all results in the first several pages are related to apple store. In this scenario, if more interaction allowed, he would feel less confused: i.e. a search engine provides guidance of concepts related to keywords already entered: i.e. fruit, company, or a clear semantic relationship structure (keyword based) of web pages retrieved that users can click on, to refine their queries.

Thus in this paper, we develop HierarSearch, which aims at enhancing performance of search engines by mining semantic relationships among results. For each user query, besides retrieving web pages, HierarSearch also generates a 3-level hierarchy aiming to provide further assistance for users. If users don't interact with the hierarchy, Hierar-Search has no difference with current general search engines. However, a hierarchy provides users choices to do further investigation if they are not satisfied with current results. Terms in the hierarchy reveal the semantic relationships within documents retrieved. The hierarchy is actually represented in the form of 3 tag clouds, where terms in one tag cloud has relationships with terms in other tag clouds. The 3 tag clouds are: concept, instance and attribute. The hierarchy can assist users in finding keywords they are interested, and the keywords can be used in further queries. To increase usability of search engines, users can click on terms in tag cloud. Terms in the tag clouds are in lexicographic order. Important terms in hierarchy graph are in bold. Users can easily select preferred keywords they want to explore further. The results will be re-ranked once they select terms in the hierarchy to refine queries. What's more, HierarSearch will not only re-rank websites based on selected terms, but also use an intelligent prediction algorithm based on random walk to "guess" user's intention behind selected terms.

Let's use an example to illustrate the process. As shown in Fig. 1, a user enters "apple" which is an ambiguous term. Besides of showing retrieved results ranked in a list like general search engines: Google, Bing, our search engine generates a 3-level hierarchy represented by tag clouds, including (1) Concept cloud: {product, reference site, news site, device, manufacturer, smart phones, company...}; (2) Instance cloud: {CNET, Iphone, iPad, macbook pro, wsj, business week...}; and (3) Attribute cloud: {story, owner, purpose...}. Once users choose "reference site" in concept level, the system firstly guesses that instance "wiki" may relate to user's intention most, since Wikipedia is one of the most influential reference websites. And then, in the tag clouds, predicted keywords are highlighted. Thus, webpages related to "wiki" of "apple" are ranked higher. In another case, if users choose "news site" in tag clouds, the algorithm in HierarSearch would assign higher scores to web pages related to "News" about "apple" when ranking.

Now we will talk about how HierarSearch generates a 3-level hierarchy. While generating 3-level hierarchy, HierarSearch needs to find real world concepts/attributes/instances by mining documents retrieved. This requires Hierarchy understands human world. Thus we desires a database, which contains the relationships among real world concepts, attributes, and instances. Since Probase [1] classifies the keywords extracted from the web and computes correlations (probabilities) between different concepts,

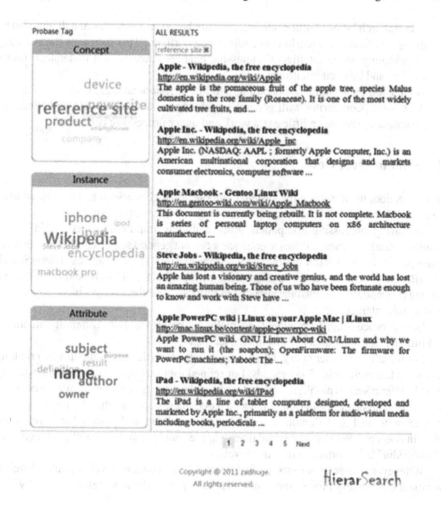

Fig. 1.

instances or attributes represented by keywords, we choose it as the knowledge base in our system. Other knowledge databases such as Freebase [2] and Cyc [3] could also be used for our system. Obviously, if knowledge in those databases can be integrated, HierarSeach can understand the world better. How to integrate multiple knowledge databases to make search more effective is left for further work.

Comparing with existing search engines, our system has following contributions.

1. The 3-level hierarchy in HierarSearch generated from results enables users to know semantic structure of results, which adds features to current list view of web pages. Users get assistance from knowing potential terms that help refine search query.

2. Interaction is allowed in HierarSearch by clicking on terms in the hierarchy. What's more, the hierarchy is represented in 3 tag clouds, which increases the usability of search engines.

3. HierarSearch helps to refine queries, not only based on keywords match, but also by guessing related real world concept/instance/attribute. This increases the accuracy of searching, since intelligently guessing enables HierarSearch understand user's explicit and implicit intentions.
4. With a sufficient offline preparation, building hierarchical tag clouds is efficiency. As a result, comparing with general search engine such as Google and Bing, our system only requires a little extra work and query processing is efficient.

2 System Overview

In this section, we introduce the system components of HierarSearch. The core components are shown in Fig. 2: document processing and query processing.

Document processing component aims at finding important terms within search results. Since HierarSearch aims at enhancing current general search engines, documents (web pages) in our search engine comes from existing search engines. HierarSearch identifies key terms in retrieved documents with the help of TF-IDF and relationships among concept/attribute/instance existed in Probase. And then 3-level hierarchy is built up for user interaction.

Query processing Component normally responds to two different situations. (1) Initially new user query (2) Refined user query.

For initially new user query, HierarSearch outputs retrieved documents in a ranking list and a hierarchy with 3 tag clouds. For refined user query, as user interacts with tag clouds, HierarSearch highlights related terms at the same time by performing a random-walk-based algorithm, and output re-ranked documents once a refined query is formed. The example in Fig. 1 demonstrates a scenario in situation 2: user selects "reference site" in concept level to refine initial query "apple", and the system intelligently guesses "Wikipedia" is important in instance level.

Whenever a user chooses specified keywords in concept level, instance level or attribute level, query processing component firstly finds the most related terms by

looking up the inverted index for the related documents. Scores of each related document are computed according to those keywords and their level weights. After scores have been calculated, documents are ranked according to them. Users can interact with HierarSearch several times and change specified keywords in different levels. If users change choices then the document processing components are invoked again to re-rank documents.

3 Demonstration

We plan to demonstration HierarSearch in understanding the principles of hierarchy search as well as the impact of its flexibility and accuracy. Since HierarSearch uses Bing to get the initialized search results and a large knowledge base extracted from billons of web pages, it can handle keywords in open fields, as common search engines. To demonstrate the benefits of our system, we have prepared a set of keywords with ambiguities such as "apple", "mouse" and spider. Users can also input any keywords for the search.

Similar as Google and Bing, our system provides an elegant search interface for users. Users can input keywords in the textbox on the web page. They do not have to find keywords accurately matching their intentions. They are only required to input the keywords related to the goal of search.

When the search button is clicked, the initialized search results are shown as three tag clouds containing concepts, instances and attributes, respectively, as well as ranked links to the documents in the web. An example is shown in Fig. 1. Cloud tag is a unique feature of our search system. The keywords in the cloud have priorities which are shown to users by the size of the keywords. Since it can be determined which keywords have closer relationships to the given query according to the knowledge base, those frequency keywords generated from the knowledge base according to the search log are bold. These bold words in the cloud can give user clues for refine the keywords.

Users could refine the search results by clicking keywords in the clouds related to their search intentions. Those clicked keywords will be used for intelligent re-ranking. Once users have chosen words in three clouds, the search engine re-ranks documents. The re-rank methodology is based on random walk. Then the refined results are shown in the interface similar as Fig. 1. During demonstration, to compare our system and traditional search engine, the comparison between Bing and HierarSearch are shown in the system.

For further interaction with the system, if users do not satisfy the search results, they can choose other keywords in clouds and re-rank results again until the satisfactory results are shown.

Acknowledgements. This paper was partially supported by National Sci-Tech Support Plan 2015BAH10F01 and NSFC grant U1509216, 61472099, 61133002. The Scientific Research Foundation for the Returned Overseas Chinese Scholars of Heilongjiang Provience LC2016026.

References

1. Probase. http://research.microsoft.com/en-us/projects/probase/
2. Bollacker, K., Evans, C., Paritosh, P., Sturge, T., Taylor, J.: Freebase: a collaboratively created graph database for structuring human knowledge. In: SIGMOD (2008)
3. Lenat, D.B., Guha, R.V.: Building Large Knowledge-Based Systems: Representation and Inference in the Cyc Project. Addison-Wesley, Reading (1989)
4. Kuang, D., Li, X., Ling, C.X.: A new search engine integrating hierarchical browsing and keyword search. In: IJCAI 2011, pp. 2464–2469 (2011)

Research on the Localization of High-Quality Teaching Resources

Deng Hong[✉], Haoliang Qi, Leilei Kong, Changwei Wu, and An Bo

Computer Science and Technology Department, Heilongjiang Institute of Technology,
Harbin 150050, Heilongjiang, China
denghdh@163.com

Abstract. The emergence of the MOOC leads to the revolution of education. All people can be taught by excellent teachers. But the interest of most of the MOOC learners only last less than two weeks. The main reasons may be the courses too difficult to study and without the restraint mechanism. This paper tries to solve the problem that learners cannot last in a MOOC. We put forward that it is feasible to bring the MOOC of elite courses to general university classroom after the localization of high quality teaching resources. We imported *Information and Entropy* of MIT to the curriculum of Heilongjiang Institute of Technology. First, we combine with own school *characteristics in Information and Entropy*. Second, we brought part of *Information and Entropy* in our class and edit of the original video in some part. Third, we have a test in the end of the course. At last all of students of class finished the instructional video and got good grades in test. The method of bringing the MOOC of elite courses to the curriculum of general university outperforms that of traditional teaching meth.

Keywords: MOOC · Localization · Instructional video

1 Introduction

2012 is the first year of the MOOC [1], MOOC which increased emphasis on informal learning are changing the way we learn. Coursera [1], one of the largest MOOC providers, reported that there were 2.8 million students registering in March 2013 [2]. And high prestige universities and courses are in Spanish, Italian and Chinese, thus showing evidence of strong user demand. In China some universities such as Shanghai Jiaotong University produce its own MOOCs and at same time import MOOCs of other universities [4].

It cannot be ignored for the simple reason that MOOC promotes sharing information worldwide and has created many opportunities for teaching and learning in a variety of disciplines [5–8]. Yet what kind of MOOCs are suitable for the students of theirs own school? It is different that different universities are requiring on the same knowledge. The better the University is, the more difficult the course is. University scholars generally believe that it is better to introduce MOOCs of the similar universities for their students to study. But in this paper, we put out that general university can import the elite courses of prestigious universities. We introduce *Information and Entropy* of MIT

© Springer Science+Business Media Singapore 2016
W. Che et al. (Eds.): ICYCSEE 2016, Part II, CCIS 624, pp. 206–210, 2016.
DOI: 10.1007/978-981-10-2098-8_24

(Massachusetts Institute of Technology) to *Information Technology* of Heilongjiang Institute of Technology.

Information and Entropy is a professional course in MIT. MIT is one of the most influential universities in the world. And it is a leading university in the world of high technology and advanced research. In the other hand, Heilongjiang Institute of Technology is a general university which cultivates the application oriented talents in the domestic. The two universities have a large gap. However, in the course of construction, we introduced some of *Information and Entropy* which is the open class of MIT to our *Information Technology*. Although many education experts think that the approach is not desirable, but the teaching method is feasible.

First, we combine with own school characteristics in *Information and Entropy*. The two courses are similar structure. Second, we introduce part of *Information and Entropy* in our class and edit of the original video in some part. The difficult of the course have been geared down. Third, we had a test at the end of the course, and deepen the restraint mechanism. At last the scores of students learning MOOC outperform the best ones of the traditional.

2   *Information and Entropy* of MIT

Information and Entropy is a demonstration class of MIT of the United States. It includes Bits and Codes, Compression, Noise and error, Probability Communications, Processes, Inference, Maximum Entropy, Physical Systems, Energy, Temperature, Quantum information. The course is for undergraduates.

The speakers of *Information and Entropy* are Paul Penfield and Seth Lloyd. Paul Penfield is the honorary professor of electrical engineering and computer science. His technical interests have included solid-state microwave devices and circuits, noise and thermodynamics, electrodynamics of moving media, circuit theory, computer-aided design, APL language extensions, integrated-circuit design automation, and computer-aided fabrication of integrated circuits. Seth Lloyd is the professor of mechanical engineering. He is the pioneer of quantum computing: he proposed the first technologically feasible design for a quantum computer. He is a Fellow of American Physical Society in 2007, and has Quantum communication, measurement, and computation prize in 2012.

Information and Entropy is imported by Classroom teaching video. The teaching language is English. Because of the differences in language, students in other countries are more difficult to learn.

3   The Structure of *Information Technology*

Information Technology is the professional course of computer science and technology institute of information management and information systems of Heilongjiang Institute of Technology. The major teaching object of it is Junior.

In 2005, *Information Technology* is been opened based on *Information and Entropy* of MIT combining with the specific practice of our school. We introduced Bits and

Codes, Compression, Probability and Maximum Entropy into the course, and joined Information Collection, Information Storage, Information release sections.

In 2009, the research group had compiled the book of *Base of Information Technology* and published by the China Water Conservancy and Hydropower Publishing House. The book [9] is composed of 8 parts. They are Information and Entropy, Information Collection, Storage Technology, Information Compression and Calibration, Information Retrieval and Utilization, Information Release, Information Security and Entropy Theory. In the same year, *Information Technology* is one of excellent courses of Heilongjiang Institute of Technology.

From now on, we have been exploring the methods of curriculum reform of *Information Technology*.

4 Introduce *Information and Entropy*

With the development of information technology and MOOC, we are also exploring the combination of flipped classroom and MOOC. We plane introduce the video of *Information and Entropy* of MIT into our course of *Information Technology*. For the huge difference of Heilongjiang Institute of Technology and MIT, we import some videos of *Information and Entropy* of MIT for our course. These are Bits and Codes, Probability, Compression, Noise and Errors, Communication, Maximum Entropy.

Information Technology takes 32 h. We divide the teaching time into 3 parts, including video, teaching and discussion. The time of videos is 7 h in all. In March 2014, we put forward the introduction of the open class curriculum *Information and Entropy* of MIT into *Information Technology*. Then we prepared the teaching plan and teaching content. And we realized the teaching reform test in the first semester of 2015 to 2014.

4.1 Class and Online Course Learning Platform

The Teaching object is Junior of information management and information systems of 2012.

The online learning platform that we chosen is ChaoXing [4] educational resources public service platform. The online learning platform includes two kinds of users. The users are students and teachers. It includes the functions such as Watch Video, Job Submission, Examining, Discussing, Progress View, in the interface of student. And it includes the functions such as Data Upload, Statistics, Homework Correcting, Upload the test questions, Reply in the interface of teachers.

4.2 Performance Comparison

Here we contrast the scores of introducing *Information and Entropy* of MIT with the scores of the traditional teaching method in the chapters of Information Theory, Information Compression and Verification, Communication and Maximum Entropy. There are 39 students in the class of 2014, and 42 students in the class of 2015. They use the same papers in *Information Technology*.

$E(X)$ is the average, $D(X)$ is the variance. The Table 1 shows that $E(X)$ of importing *Information and Entropy* of MIT are higher than that of 2014, $D(x)$ of importing *Information and Entropy* of MIT is relatively low, grade is widely improved.

Table 1. Comparison of results in teaching

	Elements of Information Theory (22)		Information Compression and Verification (24)		Communication (10)		Maximum Entropy (8)	
	E(X)	D(x)	E(X)	D(x)	E(X)	D(x)	E(X)	D(x)
2014	13.96	6.78	15.22	9.88	6.34	1.23	5.27	1.10
Importing Information and Entropy of MIT	15.23	5.26	17.61	6.27	6.93	1.09	5.54	0.70

Figure 1 is the line graph of scores distribution of the two years. It shows the score of importing Information and Entropy of MIT outperforms that of 2014.

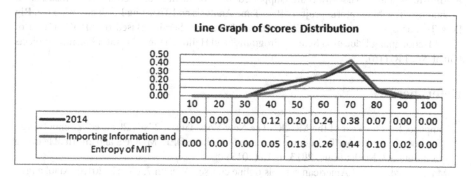

Fig. 1. Comparison of Scores Distraction (Color figure online)

In other hand in order to pass the exam, all the students insisted on finishing the video.

4.3 Discussion

Information and Entropy of MIT is teach in English, and the course is very difficult. *Information Technology* of our school is for Junior. But why did the students can accept it and understand it in our class?

First, the two courses are similar structure. The *Information Technology* of our school is based on *Information and Entropy* of MIT, and combining with own school characteristics.

Second, we edited of the original video in some part. For the differences between the two institutes, we edited the video according to the teaching plans, and adjusted the teaching content.

Third, we emphasize the difference of the course. We faced the differences between teachers and students in two schools, but we tried to weaken the difference in the classroom. So the students could study in the world's top class.

At last, we had some test in class and a final exam in the end, deepen the restraint mechanism.

5   Conclusion

It is success to introduce *Information and Entropy* of MIT to *Information Technology* of Heilongjiang Institute of Technology. That is we can introduce elite course to normal institute. From the education results, we can see students can follow the videos of MIT, finish it and the score that the students got outperforms that of traditional teaching method. There are two reasons. The one is that students worship the teachers of MIT. Introduced resources is necessary. The second is students can understand in class. In order to let students understand the introduced resources, we need to make the high-quality teaching resources localization, which reduces the learning difficulty. So it is feasible to introduce the elite courses into the classroom of the general university.

Acknowledgment.   The work are supported by Foundation of Heilongjiang Institute of Technology Teaching reform project "Real Time Analysis and Teaching Feedback Based on Big Data Technology", "Research on Personalized Education Service Based on SPOC"(GJ14) of "2014 Heilongjiang Education Science Programs" and Heilongjiang Education Planning Projects (Grant No. 14G116).

References

1. Pappano, L.: The Year of the MOOC. The New York Times, 2 Nov 2012. edinaschools.org
2. Miyazoe, T., Anderson, T.: Interaction equivalency in an OER, MOOCS and informal learning era. J. Interact. Media Educ. **2013**, 1–15 (2013)
3. Ma, L.: Awkward of American schools online courses. Wealth **7**, 48–51 (2015). Xiaokang
4. Han, M.: The 19 Universities will be Mutual recognition of Credits in Shanghai. Orient times, 21 April 2014
5. Martin, F.G.: Will massive open online courses change how we teach? Commun. ACM **55**(8), 26–28 (2012)
6. Head, K.: Massive Open Online Adventure, The Digital Campus-Chronicle of Higher Education. http://chronicle.com/article/Massive-Open-OnlineAdventure/138803/. Accessed 18 June 2013. Qi, H., Zheng, X.
7. Roth, M.: My modern experience teaching at MOOC. http://chronicle.com/article/MyModern-MOOC-Experience/138781/. Accessed 20 June 2013
8. The Digital Campus 2013, Major Players in the MOOC Universe, The Chronicle of Higher Education, June 2013. http://chronicle.com/article/The-Major-Players-in-the-MOOC/138817/. MOOC Universe
9. Qi, H.: Information Technology. China Water Conservancy and Hydropower Publishing House (2009)

Storage and Parallel Loading System Based on Mode Network for Multimode Medical Image Data

Xiao Zhai, Haiwei Pan[✉], Xiaoqin Xie, Zhiqiang Zhang, and Qilong Han

College of Computer Science and Technology,
Harbin Engineering University,
Harbin 150001, China
panhaiwei@hrbeu.edu.cn

Abstract. Since Multimode data is composed of many modes and their complex relationships, it cannot be retrieved or mined effectively by utilizing traditional analysis and processing techniques for single mode data. To address the challenges, we design and implement a graph-based storage and parallel loading system aimed at multimode medical image data. The system is a framework designed to flexibly store and rapidly load these multimode data. Specifically, the system utilizes the Mode Network to model the modes and their relationships in multimode medical image data and the graph database to store the data with a parallel loading technique.

Keywords: Multimode · Medical image data · Mode network · Graph database · Parallel loading

1 Introduction

Medical image processing attracts much attention in modern diagnostic and health-care over the past decade [1]. It can provide a lot of applications such as information retrieval and finding out more useful information that doctors are interested in. Specifically, data modelling and storage are the primary parts of the Medical image processing.

In multimode medical image data, there are many kinds of modes and complex relationships between them. For example, DICOM (Digital Imaging and Communication in Medicine) is the most common medical image file type. Besides image data, it contains many other types of information, such as hospital information, patient information, physical exam information and imaging information [2]. It is necessary to put forward new modelling and storage techniques. The techniques should integrate all modes and their relationships in multimode medical image data, so that information retrieval can obtain a more comprehensive result and data mining can be more flexible in storage structure based on the techniques.

An existing typical technology of processing medical image data is PACS (Picture Archiving and Communication Systems). It is an integrated system of digital products and technology allowing for acquisition, storage, retrieval, and display of radiographic images [3]. The system usually utilizes many tables to model the different modes of DICOM image files the file system to store DICOM image files. Because the tables are

© Springer Science+Business Media Singapore 2016
W. Che et al. (Eds.): ICYCSEE 2016, Part II, CCIS 624, pp. 211–216, 2016.
DOI: 10.1007/978-981-10-2098-8_25

independent and built for data initial demand of storage and query, the storage structure limits the multimode or cross-mode retrievals or data mining. And because the table structure is difficult to modify later, the new type of data cannot be added to the database incrementally. Moreover, the multimode or cross-mode retrievals will make the database produce a large number of table joins, which greatly reduces the query efficiency of the relational database.

Graph database is applied in areas where information about data interconnectivity or topology is more important, or as important, as the data itself. The data and relations among the data in graph database are usually at the same level [4]. The benefits of using a graph data model are given by: the introduction of a level of abstraction which allows a more natural modelling of graph data; query languages and operators for querying directly the graph structure; and ad-hoc structures and algorithms for storing and querying graphs [5]. Since all data is stored in one graph, data mining tasks cannot be limited by too many restrictions of the data storage structure. And graph database over-comes the defects that the relational database may produce a large number of table joins. So, in this paper, we utilize the graph database to store multimode medical image data.

But graph database lacks the meta information about the structure of the data stored in it. In order to identify the different modes or relationships of data in graph databases, we propose the Mode Network to model the multimode medical image data. It utilizes vertexes or edges to represent the modes and their relationships of data. Based on Mode Network and graph database, we implement a parallel loading system for multimode medical image data. The system provides the interfaces of establishing the Mode Network for uses by Web pages. And it loads the data with the parallel strategy.

2   System Architecture

The system is composed of the Client and the Server. The Client provides the data source management and Mode Network management by Web interface. Users can create a data dictionary using the data source management and build a Mode Network using the Mode Network management. Then the data dictionary and the Mode Network will be sent to the server for extracting and loading data.

The Server is mainly composed of the document database, the parallel loading system and the graph database. The document database is mainly used to store the batch files, data dictionary, Mode Network and other parameters. Parallel loading system consists of a several Import Nodes and Work Nodes as well as the version controller. The import nodes and work nodes run on a thread pool. The import nodes are responsible for parsing DICOM files and negate update transactions of graph database. The work nodes are responsible for performing these transactions. The version controller is responsible for notification all work nodes to commit the current transactions to the database periodi-cally. The graph database is responsible for response transactions committed by work nodes and other operations (Fig. 1).

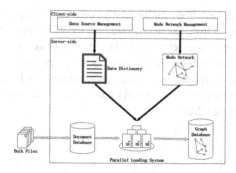

Fig. 1. System architecture

3   Methods

3.1   Establish the Mode Network

The Mode Network is built according to the requirements of users and the structure of DICOM files initially. When the requirements of the retrieval or mining is changed, or the new data is generated and needed to be loaded, the Mode Network can be modified dynamically. There are two steps in creating the Mode Network:

1. Establish the data dictionary of the DICOM file. According to the DICOM standard, the data dictionary includes Group Number, Element Number, Element Name, Element Type and Element Length. Table 1 is a fragment of Data Dictionary about the patient's information in DICOM standard.

Table 1. Fragment of DICOM data dictionary

Group Number	Element Number	Element Name
0010	0010	Patient Name
0010	0020	Patient ID
0010	0040	Patient Sex
0010	1010	Patient Age

2. Define modes and their relationships in the Mode Network. The definition of a mode includes its name, description, type, dependent modes, attributes and filter conditions. The type can be chosen from the text mode or the image mode. If the image mode is chosen, the image's inherent parameters should be specified from the mode's attributes such as the width, height. The dependent modes should be loaded before the current mode. The attributes can be chosen from the data dictionary or set as a constant. The filter conditions are also chosen from the dictionary and given regular expressions as the criterion deciding whether to load a file or not.

   The definition of the relationship between two modes includes name, description, connected modes, type, conflict-solving strategy, attributes and filter conditions.

The type can be chosen from multiple edges or single edge. The conflict-solving strategy can be chosen from the overlayable for the same attributes or the addible for new attributes. Figure 2 is a very simple example of Mode Network. It only has four modes, and three relationships among the modes. The patient mode is a text mode, its dependent mode is the hospital mode and its attributes can be chosen from Table 1. The Image is an image mode, it depends on the head mode, and its attributes can also be chosen from the data dictionary.

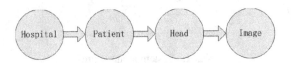

Fig. 2. Example of Mode Network

3.2 Parallel Load DICOM Files

Multiple Work Nodes create vertexes and edges in the graph database at the same time by using Read-Write lock on every vertex to avoid the interference between Work Nodes. For example, if several files contain entities of one mode that has the primary key and these files are loaded by multiple Work Nodes, every Work Node will get the Write lock of the entity's vertex before loading, so that the entities that have the same primary key will not be loaded to the database repeatedly.

1. Traverse the import directory that the user set and generate a file list to be loaded.
2. Traverse the Mode Network, generate the modes' loading order based on whether a mode is dependent on another mode and whether a mode is connected to an input edge, then inform all Import Nodes of the loading order.
3. An Import Node gets one file from the file list each time. Then according to the data dictionary, Mode Network and the loading order, the Import Node extracts data from files, creates graph update transactions and sends the transactions' number to the queue of the version controller. Finally, all transactions generated by one file are assigned to one Work Node.
4. When the transaction queue of a Work Node is not empty, the Work Node executes sequentially.
5. The version controller takes a transaction number from its queue periodically and notify all the Work Nodes. After receiving the notification, all Work Nodes submit the transactions whose numbers is less than the received number.

4   Demonstration

To demonstrate the running process and efficiency of the system, we selected four datasets of DICOM files for loading. The files are about real CT images of patients' head in a hospital. The computer's memory is 16G and the processor is Intel Core i7 2.2 GHz.

In the data source management interface, we create an empty data dictionary firstly, then add dictionary entries to the dictionary based on DICOM standard. And in the Mode Network management interface, we build an empty Mode Network firstly. Then we enter the mode management interface and create modes. We created four modes, Hospital, Patient, Head and Image shown in Fig. 2. And we enter the mode relationship management interface and create mode relationships shown in Fig. 2. We created three ownerships, from Hospital to Patient, from Patient to Head and from Head to Image.

After loading a batch of DICOM files, we can view the data stored in the graph database by visualization, as shown in Fig. 3. The yellow vertex in the Fig. 3 represents an entity of the Hospital mode, two red vertexes represent entities of Patient, the two center vertexes of the two groups of vertexes represent entities of Head, and other vertexes around them represent entities of Image mode. And the direction edges in Fig. 3 represent the modes' relationship. Because Fig. 3 is only a small part of the whole graph, there is just a fraction of fully loading result. Figure 4 shows the running time. We can see the system gets the maximum efficiency when the number of thread pool is equal to the number of processor's core.

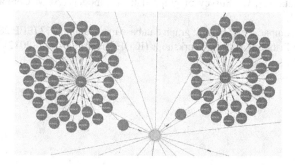

Fig. 3. Loading result (Color figure online)

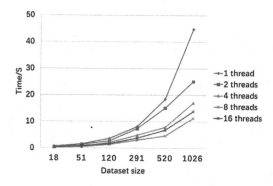

Fig. 4. Loading result (Color figure online)

5 Conclusion

This paper presents graph-based storage and parallel loading techniques for multimode medical image data and implements a system. The system integrates multiple modes of data by the Mode Network and the graph database. Therefore, the system supports efficient multimode or cross-mode retrieval, and reduces the limitations from the storage structure. Next, we will focus on information retrieval and data mining in the system.

References

1. Liu, Z., Yin, H., Chai, Y., Yang, S.X.: A novel approach for multimodal medical image fusion. Expert Syst. Appl. **41**(16), 7425–7435 (2014)
2. Mildenberger, P., Eichelberg, M., Martin, E.: Introduction to the DICOM standard. Eur. Radiol. **12**(12), 920–927 (2002)
3. Hu, W.: Picture archiving and communication system. Comput. Tomogr. Theory Appl. **33**(1), 257–261 (2003)
4. Angles, R., Gutierrez, C.: Survey of graph database models. ACM Comput. Surv. **40**(1), 178–187 (2008)
5. Angles, R.: A comparison of current graph database models. In: 2012 IEEE 28th International Conference on Data Engineering Workshops (ICDEW), pp. 171–177 (2012)

The BBC News Hunter: A Novel Crawler for BBC News

Mingxin Wang[1], Ning Wang[1], Boran Wang[1], Can Tian[1], Yanchun Liang[2,3], Guozhong Zhao[4], and Xiaosong Han[2,4(✉)]

[1] College of Software, Jilin University, Changchun 130012, China
[2] Key Laboratory for Symbol Computation and Knowledge Engineering of National Education Ministry, College of Computer Science and Technology, Jilin University, Changchun 130012, China
hanxiaosong@jlu.edu.cn
[3] Zhuhai Laboratory of Key Laboratory for Symbol Computation and Knowledge Engineering of Ministry of Education, Zhuhai College of Jilin University, Zhuhai 519041, China
[4] Daqing Oilfield Personnel Development Institute, CNPC, Daqing 163000, China

Abstract. In order to distinguish and extract the topic information from other interferential information on the BBC news website for the study in social computing, the BBC News Hunter was proposed in this paper. The whole system consists of 6 subsystems, respectively named: UI, Control, Download, Analysis, Storage and Log. Numerical experiments show that satisfactory results can be obtained from the BBC news website, whose average accuracy as well as efficiency are acceptable.

Keywords: BBC · Crawler · News · HTML parser · Multithread

1 Introduction

News is an important way for us to know and understand the social, and it is also a kind of indispensable study and entertainment material. With the rapid development of the Internet in recent years, the spread of information is faster and more diverse, which has become the focus of the scientific community, the business community and even the governments around the world. On the other hand, getting information becomes more and more difficult in manual way because of the explosive growth of the Internet information. More and more webpages are filled with useless contents, such as spam, ads and so on, which seriously affect the search engines' effect. Therefore, how to distinguish and extract the topic information from other interferential information needs to be considered. As social computing is becoming increasingly important, the extraction of news information related to the Internet has become an important branch of the field of Information extraction area. Web crawler is a tool to provide a system to (of browsing) browse the Internet and the World Wide Web, which can perform the usual pages indexed. Since it can automatically crawl the Web information, more and more people uses it as an important way of gathering information. BBC World News website, as one of the world's most renowned international news websites, is most authoritative. Considering the timeliness of the information nowadays as well as the application of social computing, how to effectively extract and analyze the content of BBC website

© Springer Science+Business Media Singapore 2016
W. Che et al. (Eds.): ICYCSEE 2016, Part II, CCIS 624, pp. 217–225, 2016.
DOI: 10.1007/978-981-10-2098-8_26

has become an important problem. At the same time, how to research and develop a crawling and analysis tool to extract information from BBC news website has become the focus of this study.

Web crawler's basic idea is to download the pages from the Internet and extract the information we need after the analysis. We believe that an efficient crawler should have the following three features:

1. Make full use of network and CPU performance;
2. Accurately extract the required information from the web pages, and store it as the user's desired format;
3. Give Valuation and classification of web pages, valuation means analyzing download priority through URL format, classification means starting a different analysis program responding to different page structure.

In this paper, we designed and realized a system, the BBC News Hunter, which can analysis and extract the information from BBC news website. Firstly, we analyzed and discussed the key technologies of web crawler including the design process. Then we focused on the process of the implementation considerations, comparing this design with other tools. Finally, the experimental data proved the validity of our tool. Organizational structure and contents of the chapters are as follows:

The second part mainly introduces recently web crawler search engines in the development process and the technology of web analytics; the third part will describe the BBC News Hunter implementation details; the fourth part will give the performance and efficiency comparison of this design and some similar crawlers; the fifth part will summarize the full text.

2   Related Work

To obtain the information of news web pages, we need the support of crawler technology, then analyze and calculate the content of news web page. So page analysis technology is an integral part of the information collection. The network delay has become the bottleneck of the whole system performance in the process of downloading web pages. To improve the efficiency, buffer queue and multi-threaded parallel structure is designed in the crawler.

(a) The Technology of Web Crawler

A crawler is a program that automatically collects web pages to create a local index and/ or a local collection of web pages. Roughly, a crawler starts off with an initial set of URLs, called seed URLs. It firstly retrieves the pages identified by the seed URLs, extracts any URLs in the pages, and adds some new URLs to a queue of URLs to be scanned. Then the crawler gets URLs from the queue (in some order), and repeats the process [1].

Some popular crawler tools are as follows.

(1) Nutch [2] is an open-source Web search engine that can be used at global, local, and even personal scale. Nutch originated with Doug Cutting, creator of both Lucene and Hadoop, and Mike Cafarella. One of its signature features is the ability to "explain"

its result rankings. Nutch is a complete open-source Web search engine package that aims to index the World Wide Web as effectively as commercial search services [3]. But the disadvantage is that using Nutch in extracting data will waste a lot of time on unnecessary calculations. The working of Nutch relies on Hadoop. But it will consume a lot of time. If the number of clusters of machines is small, it is not as fast as the crawler working in one computer [4].

(2) Heritrix [5] is the Internet Archive's open-source, extensible, web-scale, archival-quality web crawler project. The Internet Archive started Heritrix development in 2003. The intention was to develop a crawler for the specific purpose of archiving websites and to support multiple different use cases including focused and broad crawling. The software is open source to encourage collaboration and joint development across institutions with similar needs. A pluggable, extensible architecture facilitates customization and outside contribution. Now, after many years' development, the Internet Archive and other institutions are using Heritrix to perform focused and increasingly broad crawls. But the main limitations are [6]:

(i) Single instance only: cannot coordinate crawling amongst multiple Heritrix instances whether all instances are run on a single machine or spread across multiple machines.

(ii) Requires sophisticated operator tuning to run large crawls within machine resource limits.

(iii) Only officially supported and tested on Linux.

(iv) Each crawl is independent, without support for scheduled revisits to areas of interest or incremental archival of changed material.

(v) Limited ability to recover from in-crawl hardware/system failure.

(3) Open source Crawler4j [7] is a crawler with a variety of functions. It can collect large amounts of data by widening the scope of the search like General crawlers and Focused crawlers, and anointing a target for accurate search just like topical crawlers. Accurate retrieval can also be appointed for the crawlers to search. However, the problem is that it uses a lot of memory due to the excessive use of global variables, and unnecessary duplication occurs in collection. In addition, it is difficult to modify, since the amount of crawler source is massive and there are also frequent errors and stuttering occurrences.

(b) The Technology of Web Analysis

Web analysis is a hot topic in recent research. In the middle of 1960s, the research of structured information from natural language text was regarded as the initial research of information extraction technology. In recent years, with the popularity of the Internet, information in the form of HTML page has become more and more prevail on the Internet. Information extraction researchers also began to gradually pay attention to this aspect, so many HTML page information extraction algorithm has been proposed. At present, the method of web page information extraction based on page structure analysis mainly includes the following methods [8].

(1) DOM (Object Model Document) tree. Because DOM tree directly displays the structure of HTML page, we can use the DOM tree to analyze the structure of the page

layout [9]. A method about web information extraction based on Ajax is proposed through the thorough investigation into the traditional Web information extraction, which can better extract information of dynamic web pages [10].

(2) Visual features of the page. Combining the location information of the object and the spatial relationship, we can better understand the web pages, and identify the subject information. For example, Kovacevic et al. [11] in 2002 constructed a M-tree using visual information and then define rules to identify common page title area, left and right menu, footer and page center; Kang and Choi [12] in 2007 based on VIPs page blocking algorithm proposed the RIPB algorithm (Recognizing Informative Page Blocks) in order to find the subject information block. It analyzed web page by the algorithm of Visual Web page segmentation.

(3) Distribution of HTML markers. Embley [13], Brigbham Young University, proposed five heuristic rules, according to the key words from the processing problem. The boundaries of records could be established by these rules. Considering the problem that some pages lack common features, this method may fail. China University of Petroleum (East China) Suo [14] proposed a method based on the web content extraction of tag window. Through regular expressions, text between labels can be extracted. And the realization of this method is simple. The disadvantage of this method is that the implicit semantic information of the web page is not considered.

Therefore, to improve the efficiency and accuracy of crawling, we proposed a system, the BBC News Hunter, which can analysis and extract the information from BBC news website, in order to meet the demand for accessing to useful news.

3   The BBC News Hunter

(a) Function Description

The BBC News Hunter consists of 6 subsystems, named: UI, Control, Download, Analysis, Storage and Log, respectively. The overall structure of the system is as Fig. 1.

The UI subsystem is designed for users to control the resources in the entire system, occupying a free-running thread. The Control subsystem maintains all the key information in the system, real-time scheduling according to the running condition of the system. The Download subsystem downloads web pages from the Internet according to the URLs given, handling the exceptions runtime. The analysis subsystem provides a complete solution to analyze a target web page, converting web pages to data ordered in some specific structure. The Storage subsystem writes the valid data to the disk. The Log subsystem records the trajectory of the whole system.

(b) Implementation Details
(1) Thread Management

The design of "BBC News Hunter" is based on multithread model. Thus it's necessary to manage the threads in a way that should promise the performance and the consistency of the system. We employed the thread-pool method and designed a thread-level manager.

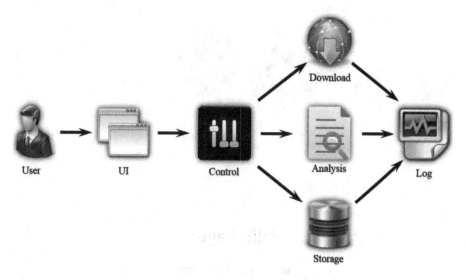

Fig. 1. Structure of the BBC News Hunter

(2) Crawling Process

All the worker threads in the thread-pool have the same running process. Worker threads will firstly apply a new crawling task from the Control subsystem after the initiation process, then perform a crawling cycle with the cooperation of the three subsystems named Download, Analysis and Storage as Fig. 2.

(3) Valid Pages Screening Strategy

A valid page is either a topic page or a news page. From a topic page, we can extract some related hyperlinks. From a news page, not only can we obtain some related

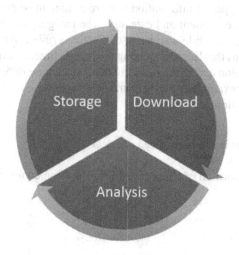

Fig. 2. Crawling process

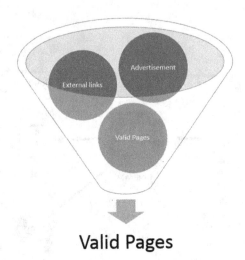

Fig. 3. Screening strategy

hyperlinks but a piece of valid news. By analyzing, we found that all valid pages have a URL, ignoring the common hostname, which strictly meets the regular expression: "^([a-z]+-)+[0-9]+$", representing that a URL should consist of several English words and end up with a decimal number, spreading with '-'s. For example: "uk-england-10884539" and "world-32980451" are valid URLs, while "also_in_the_news" and "contact" are not. So we determine the validity of a URL by prejudging whether the one meets the regular expression above. Figure 3 shows the process of Screening Strategy.

(4) Analysis Strategy

Because of the variety of the structure of the news pages, we designed an analysis strategy. There are 4 types of information that are needed to be extracted from a news page: title, release time, content and category. The category can be simply extracted from the URL format. The title also has a distinctive symbol. An area exists in each news page that contains the data representing the release time. Considering there is a lot of redundant information across the news content, we can use different tags in order to filter information. For example as Fig. 4, from URL "business-34277739" can we extract the following information:

(1) Date: 2015-09-17
(2) Title: China firms in US high-speed rail deal
(3) Category: Business
(4) Content: "A consortium of Chinese rail firms has teamed up with private US …"

Fig. 4. Web page for URL "business-34277739"

(5) Data Storage

In order to simply maintain and manage the data extracted from the BBC website, we stored the data in file format and establish a multi-level directory storage for different day and different categories of information. For example, we store the extracted data from URL "business-34277739" to folder "info\2015-09-17\" in txt file format.

4 Experimental Data

At this stage, we tested four crawler tools Nutch2.3, Heritrix3.0, Crawler4j, and the BBC News Hunter to crawl specific information from BBC News. The first three tools are open-source tools based on JAVA, and the last one is a tool written in C++ by us. Technically, all four tools can crawl webpages' source codes and store them in hard disk. However, compared to the first three, which can only simply download webpages' source codes, the BBC News Hunter has an evident advantage: analyzing web property and grabbing headlines, posting time, and content directly. To have a better comparison, we run these tools to crawl information from BBC News website in the same one-hour-long period under the same environment: 50 threading, default priority, and 100 Mbps network bandwidth. The experimental results are presented at the bar chart as Fig. 5.

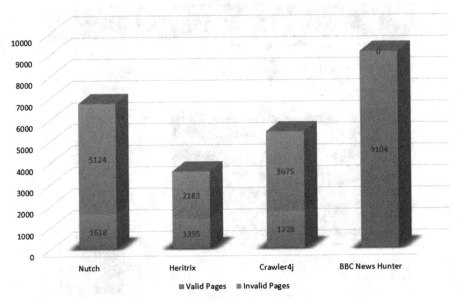

Fig. 5. Experiments results (Color figure online)

As shown in the figure, the performance of our crawler is more excellent than other tools in speed and efficiency, and our tool can directly extract web information from web page source code, while the other three tools can only extract the web page source code. Table 1 shows the details.

Table 1. Experiments results

Crawler name	Total pages	Valid pages	Valid rate	Result type
Nutch	6742	1618	24 %	HTML
Heritrix	3578	1395	39 %	HTML
Crawler4j	5403	1728	32 %	HTML
BBC News Hunter	9104	9104	100 %	Pure Text

5   Conclusion

The development of Internet boosts the revolution of news communication, and the advancement of search technology is pushed by retrieval method of the Internet information. What's more, the crawler technology and web page analysis make a tremendous contribution to the collection of news information.

This paper was built around crawler designs and webpage's analytical techniques, which promoted the development of social computing. Specifically, we tested and analyzed the BBC News Hunter, a self-developed software devoted to crawl information from BBC News website. The results indicated that this tool could effectively solve the inefficient reading caused by garbage in news sites, grab useful information from necessary websites, and facilitate reading and sorting. However, this tool still has limitations that it cannot crawl pictures in information collection and expand news sites. Nevertheless, we are glad to perfect these functions and apply it to other websites.

Acknowledgements. The work was supported by National Science Foundation of China under Grant 61503150, 61472158, 61572228 and the 2015 Annual Innovation Training Program of Jilin University under Grant 2015540784.

References

1. Wang, J., Zhu, L., Li, C.: Discussion about the core of search engine again—web crawler. In: 2011 International Conference on Computer Science and Service System (CSSS), pp. 3188–3191. IEEE (2011)
2. Khare, R., Cutting, D., Sitaker, K., Rifkin, A.: Nutch: a flexible and scalable open-source web search engine. Or. State Univ. **1**, 32 (2004)
3. Brin, S., Page, L.: Reprint of: the anatomy of a large-scale hypertextual web search engine. Comput. Netw. **56**(18), 3825–3833 (2012)
4. http://blog.csdn.net/chaishen10000/article/details/50776662
5. Mohr, G., Stack, M., Ranitovic, I., et al.: An Introduction to Heritrix An open source archival quality web crawler. In: IWAW 2004, 4th International Web Archiving Workshop (2004)
6. Liu, D.F., Fan, X.S.: Study and application of web crawler algorithm based on heritrix. In: Advanced Materials Research, vol. 219, pp. 1069–1072. Trans Tech Publications (2011)
7. Kim, H.G., Lee, J.W., Ban, T.H., Jung, H.K.: A study on distributed crawling-based overhead optimization. Int. J. Softw. Eng. Appl. **9**(3), 175–182 (2015)
8. Feng, W., Mao, Z.: The research of web pages information extraction based on Web. J. Luoyang Technol. Coll. **3**, 30–31 (2005)
9. Chakrabarti, S.: Integrating the document object model with hyperlinks for enhanced topic distillation and information extraction. In: Proceedings of the 10th International Conference on World Wide Web, pp. 211–220. ACM (2001)
10. Hengru, Z., Chun, C.: Web information extraction technology research based on ajax. In: 2011 International Conference on Business Computing and Global Informatization (BCGIN), pp. 208–211. IEEE (2011)
11. Kovacevic, M., Diligenti, M., Gori, M., Milutinovic, V.: Recognition of common areas in a web page using visual information: a possible application in a page classification. In: Proceedings of the 2002 IEEE International Conference on Data Mining, ICDM 2003, pp. 250–257. IEEE (2002)
12. Kang, J., Choi, J.: Detecting informative web page blocks for efficient information extraction using visual block segmentation. In: International Symposium on Information Technology Convergence, ISITC 2007, pp. 306–310. IEEE (2007)
13. Embley, D.W., Jiang, Y., Ng, Y.K.: Record-boundary discovery in web documents. ACM SIGMOD Rec. **28**(2), 467–478 (1999). ACM
14. Zhao, X.X., Suo, H.G., Liu, Y.S.: Web content information extraction method based on tag window. Jisuanji Yingyong Yanjiu/Appl. Res. Comput. **24**(3), 144–145 (2007)

Time-Based Microblog Search System

Zhongyuan Han[1]([✉]), Wenhao Qiao[2], Shuo Cui[3], and Leilei Kong[1]

[1] School of Computer Science and Technology, Heilongjiang Institute of Technology,
Harbin 150050, China
hanzhongyuan@gmail.com
[2] Beijing Institute of Surveying and Mapping, Beijing 100038, China
[3] Huarun Group, Shenzhen 518001, China

Abstract. This demo shows a time-based microblog research system which developed based on the time profile to estimate the query model, the document model and rank function for microblog search. The system exploits the time profile to boost the performance of microblog search. A brief description of the time-based query model, time-based document model and time-based similarity score is introduced. The index strategy for temporal microblog search is described. Using TREC 2011 and TREC 2012 microblog retrieval collection, the examples of microblog search results are demonstrated.

Keywords: Microblog search · Query model · Document model · Learing to rank · Time

1 Introduction

Short texts, such as microblogs, are prevailing with the development of mobile Internet and social media. In face of the massive microblogs and the users' diverse information requirements, microblog search and microblog filtering have become the indispensable components of the microblog service. In recent years, the time profile of microblog attracts the attention of researchers. Recent works reveal that the time profile is promising in improving the performance of microblog retrieval.

This time-based microblog research system is focused on boosting the performance of microblog search by using the time profile of microblog in three research issues: i.e. query modeling, document modeling, relevance estimation between query and documents. We attempt to leverage the time profile to reduce the impacts of short texts on content-based microblog search.

2 System Overview

The system flowchart is shown in Fig. 1. The system has three core component: a Term Time Distribution-based Query Model (TTDQM), a novel document expansion model named Time-base Document Model (TBDM), and a Time-Sensitive Learning to Rank algorithm (TSL2R) to learn rank function.

© Springer Science+Business Media Singapore 2016
W. Che et al. (Eds.): ICYCSEE 2016, Part II, CCIS 624, pp. 226–228, 2016.
DOI: 10.1007/978-981-10-2098-8_27

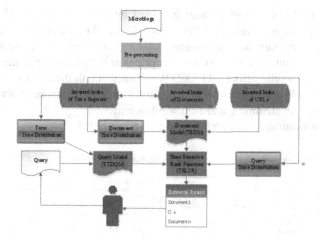

Fig. 1. The system flowchart

In TTDQM, a query model based on term time distribution is proposed to deal with the short query in microblog search. First, we define the term time distribution and analyze the temporal distribution of the queries and the relevant documents. Then a similarity measure based on Minkowski distance for term time distribution is presented to properly decide the relevant terms for an expanded query model. The suggested method t avoids the drawbacks of the classical content based query expansion approaches caused by the length limit in microblog by attempting to employ only time profile to establish the relevance between query terms and expanded terms [1].

In TBDM, a time-based document model is presented to enrich the short texts of microblogs. First, candidate terms are jointly weighted by their temporal closeness to the microblog and their distribution over different periods. Then, terms not in the original microblog are refined by a machine learning mechanism trained on pseudo-supervised data. Finally, a document expansion model is designed, together with two approximate solutions to optimize the time complexity to reduce the time cost. The content of webpage linked in tweets is also introduced into document model [2].

In TSL2R, a temporal relevance is investigated to enhance the relevance estimation of microblog search, as an additional information besides the well-recognized content relevance. First, three aspects of time relevance are explored in the framework of language model. Then, in the current framework of learning to rank, a loss function based on time-sensitive learning to rank is defined, which is aimed at a ranking more consistent with the time profile of microblogs [3, 4]. The relevant microblogs posted after the query time are provided via a hybrid filter [5, 6].

3   Demonstration

We conduct the demonstrations on the dataset of TREC 2012 Microblog Research Task. The corpus is comprised of 2 weeks of tweets sampled from Twitter. We downloaded

10,397,336 tweets by twitter crawler provided by track organizers. The main pre-processing steps are as follows: (1) Retweets without "RT" are removed since they are to be judged as non-relevant. The retweets with "RT" are removed if there is no contents in front of "RT". But once there exist descriptions at the beginning of their tweet text, we only keep the words before "RT". (2) We filtered out all the non-English tweets by using language identifier tool provided by Nutch (http://nutch.apache.org). (3) Porter stemmer is used for stemming and stop words are filtered.

The demonstrations of research results are shown in Fig. 2.

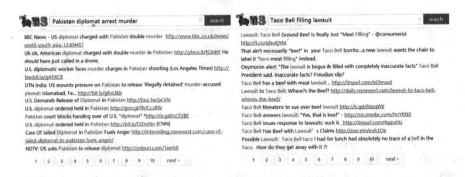

Fig. 2. Demonstrations of research results

Acknowledgments. This work is supported by Research Project of Heilongjiang Provincial Department of Education (No. 12541677).

References

1. Han, Z., Yang, M., Kong, L., Qi, H., Li, S.: Query expansion based on term time distribution for microblog retrieval. Chin. J. Comput. (2016)
2. Han, Z., Yang, M., Kong, L., Qi, H., Li, S.: A hyperlink-extended language model for microblog retrieval. Int. J. Database Theory Appl. **8**(6), 89–100 (2015)
3. Han, Z., Li, X., Yang, M., Qi, H., Li, S.: Feature analysis in microblog retrieval based on learning to rank. In: Zhou, G., Li, J., Zhao, D., Feng, Y. (eds.) NLPCC 2013. CCIS, vol. 400, pp. 410–416. Springer, Heidelberg (2013)
4. Han, Z., Li, X., Yang, M., Qi, H., Li, S., Zhao, T.: Hit at TREC 2012 microblog track. In: Proceedings of Text Retrieval Conference (2012). 3
5. Han, Z., Yang, M., Kong, L., Qi, H., Li, S.: A temporal microblog filtering model. Int. J. Grid Distrib. Comput. **9**(1), 89–98 (2016)
6. Han, Z., Yang, M., Kong, L., Qi, H., Li, S.: A hybrid model for microblog real-time filtering. Chinese J. Electron. **25**(3), 432–440 (2016)

Traffic Collection and Analysis System

Jinlai Liu, Haitao Wen, Xiangyu Hou, Guanglu Sun[✉], and Suxia Zhu

School of Computer Science and Technology,
Research Center of Information Security and Intelligent Technology,
Harbin University of Science and Technology, Harbin, China
guanglu_sun@163.com

Abstract. This paper designs and implements a network traffic collection and analysis system (NTCA). Network traffic is collected from multiple clients. According to the offline traffic analysis, network flows are analyzed and annotated with application protocol. At the same time, the results are stored in the system. NTCA also provides a visual interface to simplify the configuration process. With good flexibility and scalability, NTCA can be extended to a network traffic classification system, which is crucial to the research of network security technology.

Keywords: Traffic classification · Traffic collection and analysis · Feature extraction · Traffic annotation

1 Introduction

Network traffic classification has been widely used in network management and network monitoring in recent years. However, it's extremely difficult to collect annotated network traffic data in research, because of the uncertainty and diversity of the large amount of data.

The annotated network traffic is obtained in the following three ways. The first is artificial annotation, such as public network flow data set provided by Moore in 2005. Moore data set provided a support mostly for the machine learning based method [1]. The second is annotated by the IANA registration port or known payload information. Because only registration port and unencrypted payload can be detected, this method has much limitations [2, 3]. The third is an application layer annotation mechanism, which is used in monitoring the client core to get ground truth data [4]. But its configuration process is too complex, so that inexperienced user may encounter difficulties during configuration.

This research is partly supported by the Ministry of Education's Humanities and Social Science Project No. 11YJC740048, Scientific planning issues of education in Heilongjiang Province No. GBC1211062, research fund for the program of new century excellent talents in Heilongjiang provincial university No. 1155-ncet-008 and the National Natural Science Foundation of China under grant No. 60903083, 61502123.

W. Che et al. (Eds.): ICYCSEE 2016, Part II, CCIS 624, pp. 229–234, 2016.
DOI: 10.1007/978-981-10-2098-8_28

A network traffic collection and annotation system is designed and implemented that provides the ground truth data to support network traffic classification. Traffic collection module collects the terminal client's traffic information and the corresponding application category information [5]. Then the acquired traffic information is stored in the data storage server. Meanwhile, Network traffic analysis is provided in traffic analysis module. In this module, feature extraction is carried out for the stored traffic data. The offline data is combined with the annotated traffic, which can be applied to network traffic classification. In addition, the system can be utilized to the network for traffic classification. Finally, the system provides a visual interface which can simplify the configuration process, and gives a visual interface for the collected data in order to facilitate management.

The rest of the paper is organized as follows. Section 2 presents the detail of the system implementation. Our experimental platform is described in Sect. 3. The paper is finally summarized in Sect. 4.

2 Implementation

NTCA system is mainly divided into three parts: the traffic collection module, the data storage module, the traffic analysis module. System architecture is shown in Fig. 1.

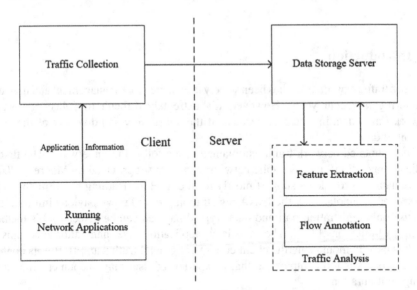

Fig. 1. System architecture

The main function of each module is described as follows:

1. Traffic collection module collects network traffic data and the corresponding traffic applications information. Network traffic data is collected by an improved capture tool, stored as pcap format. Flows are defined and differed based on 5-tuple. The

flows are meanwhile monitored by the NTCA client in order to build the relation between flows and applications in the client machine. Based on this collection module, NTCA can easily implement network traffic monitoring.

2. Data storage module includes two parts. Database storage creates a data storage server for the traffic information collection. Flow profile storage creates a file server to store the pcap file. To facilitate the transfer, the flow profiles are divided into small parts size of 10 M.

3. Traffic analysis module includes two parts. Feature extraction analyzes the packets and extracts 42 statistics feature of the flow [6, 7]. Traffic annotation labels each flow based on the relation from traffic collection module. The above information is finally combined in the module, which give a data support for network traffic classification.

Furthermore, the deployment and configuration of system are realized by the visualization platform, which provides an interface for client and server respectively. Different users have different privileges and get access to various webpage. Normal users get access to the client through the browser to install and deploy client tool. Administrator monitors and controls the client in real-time and observes the collected information. Then it acquires and analyzes the traffic information periodically.

3   Experiment

The topology of the test platform is shown in Fig. 2. The client programs run in distributed PCs complete traffic collection. Data storage server provides the data storage. Finally, the analysis server accomplishes feature extraction, flow Annotation and traffic classification.

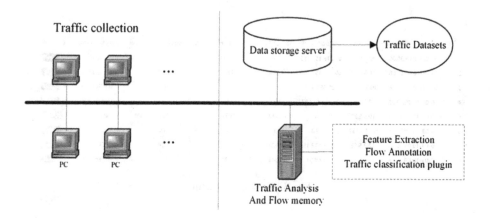

Fig. 2. The topology of the test platform

NTCA system captures all the clients at the same time. When the PCs browse web, watch online videos, download, and other network operations, the traffic information will be collected and saved in the remote storage server. Finally, the analysis module carries out feature extraction and flow annotation operations. The flow statistical characteristics are shown in Table 1 based on feature extraction. The characteristics and corresponding categories of network traffic in NTCA are shown in Figs. 3 and 4.

Table 1. Table of flow statistical characteristics.

Network traffic features		
Source IP	min_data_wire	Traffic recognition time
Destination IP	mean_data_wire	Network layer protocol
Source Port	max_data_wire	Data link layer protocol
Destination port	min_data_ip	Transport layer protocol
Application Name	mean_data_ip	Session layer protocol
Waiting time	max_data_ip	Ttl_stream_length
Duration	actual_data_pkts	The number of bytes contained
min_IAT	pure_acks_sent	Minimum TCP receive window
Zero_win_adv	Urgent_data_pkts	The method to recognize traffic
max_IAT	ack_pkts_sent	The number of packets contained
Alert level	FIN_pkts_sent	The mean length of the IP packet header
SYN_pkts_sent	Urgent_data_bytes	Maximum TCP receive window
Average TCP receive window		The minimum length of the IP packet header
Whether there is data loss		The maximum length of the IP packet header
The number of TCP packet which containing RST		

ID	traffic recognition time	source IP	destination IP	source Port	destination port	transport layer protocol	Application Name	Operation
1	2016-03-10 09:30:08	10.3.5.30	60.219.151.155	12580	10193	UDP	QQ	Detail
2	2016-03-10 09:30:08	10.3.5.30	60.219.151.155	12580	57511	UDP	QQ	Detail
3	2016-03-10 09:30:08	10.3.5.133	125.39.54.71	3474	443	TCP	WeChat.exe	Detail
4	2016-03-10 09:30:08	10.3.5.65	220.181.112.244	3480	443	TCP	chrome.exe	Detail
5	2016-03-10 09:30:08	10.3.5.32	65.52.0.51	3532	5671	TCP	devenv.exe	Detail
6	2016-03-10 09:30:08	10.3.5.133	0.0.0.0	137	0	UDP	System	Detail
7	2016-03-10 09:30:08	10.3.5.65	182.254.12.151	3628	80	TCP	QQPCPatch.exe	Detail
8	2016-03-10 09:30:08	10.3.5.52	65.52.0.51	3648	5671	TCP	devenv.exe	Detail

Fig. 3. Traffic information

Current traffic information

ID	ctime	protocol2	protocol3
1	2016-03-10 09:30:08	enthernet	ip
srcIP	**destIP**	**protocol4**	**srcport**
10.3.5.30	60.219.151.155	UDP	12580
destport	**protocol5**	**packets**	**bytes**
10193	UNKNOWN	1266	1079642
method	**alarm_level**	**min_timeinterval**	**max_timeinterval**
OTHERS	normal	0	2
min_eth_len	**max_eth_len**	**average_eth_len**	**min_ip_len**
72	1450	866	58
max_ip_len	**average_ip_len**	**min_iphdr_len**	**max_iphdr_len**
1436	852	20	20
average_iphdr_len	**ack_num**	**only_ack_num**	**leastone_ack_num**
20	0	0	0
psh_num	**syn_num**	**fin_num**	**urg_num**
0	0	0	0
rst_num	**urg_byte_char**	**min_win**	**max_win**
0	0	0	0
average_win	**zero_win_num**	**ether_length**	**miss_data**
0	1266	1097366	1
continuous	**time_spend_idle**	**Application Name**	
153	0	QQ	

Fig. 4. The diagram of the raw traffic data

4 Summary

The proposed network traffic collection and analysis system plays an important role in network security monitoring, illegal traffic controlling, bandwidth management and other network applications. NTCA collects and monitors multi-clients' traffic at the same time. Simultaneously it is capable of analyzing network traffic, extracting flow features, annotating traffic based on application category. The characteristics and annotation results which support to network traffic classification were finally stored in the storage server. The visual interface is used to manage all the modules of NTCA.

References

1. Moore, A., Zuev, D., Crogan, M.: Discriminators for use in flow-based classification. Queen Mary and Westfield College, Department of Computer Science (2005)
2. Moore, A.W., Papagiannaki, K.: Toward the accurate identification of network applications. In: Dovrolis, C. (ed.) PAM 2005. LNCS, vol. 3431, pp. 41–54. Springer, Heidelberg (2005)

3. Sen, S., Spatscheck O., Wang D.: Accurate, scalable in-network identification of p2p traffic using application signatures. In: Proceedings of the 13th International Conference on World Wide Web. ACM, pp. 512–521 (2004)
4. Gringoli, F., Salgarelli, L., Dusi, M., et al.: Gt: picking up the truth from the ground for internet traffic. ACM SIGCOMM Comput. Commun. Rev. **39**(5), 12–18 (2009)
5. Keys, K., et al.: The architecture of CoralReef: an internet traffic monitoring software suite. In: PAM (2001)
6. Moore, A.W., Zuev, D.: Internet traffic classification using bayesian analysis techniques. ACM SIGMETRICS Perform. Eval. Rev. **33**(1), 50–60 (2005). ACM
7. Kim, H., Claffy, K.C., Fomenkov, M., et al.: Internet traffic classification demystified: myths, caveats, and the best practices. In: Proceedings of the 2008 ACM CoNEXT Conference. ACM, p. 11 (2008)

Author Index

Printed in the United States
By Bookmasters